Stefan Neppl

Attosecond Photoemission from Surfaces and Interfaces

Stefan Neppl

Attosecond Photoemission from Surfaces and Interfaces

Time-resolved investigation of ultrafast electron dynamics in condensed matter with attosecond resolution

Südwestdeutscher Verlag für Hochschulschriften

Impressum/Imprint (nur für Deutschland/only for Germany)
Bibliografische Information der Deutschen Nationalbibliothek: Die Deutsche Nationalbibliothek verzeichnet diese Publikation in der Deutschen Nationalbibliografie; detaillierte bibliografische Daten sind im Internet über http://dnb.d-nb.de abrufbar.
Alle in diesem Buch genannten Marken und Produktnamen unterliegen warenzeichen-, marken- oder patentrechtlichem Schutz bzw. sind Warenzeichen oder eingetragene Warenzeichen der jeweiligen Inhaber. Die Wiedergabe von Marken, Produktnamen, Gebrauchsnamen, Handelsnamen, Warenbezeichnungen u.s.w. in diesem Werk berechtigt auch ohne besondere Kennzeichnung nicht zu der Annahme, dass solche Namen im Sinne der Warenzeichen- und Markenschutzgesetzgebung als frei zu betrachten wären und daher von jedermann benutzt werden dürften.

Coverbild: www.ingimage.com

Verlag: Südwestdeutscher Verlag für Hochschulschriften GmbH & Co. KG
Heinrich-Böcking-Str. 6-8, 66121 Saarbrücken, Deutschland
Telefon +49 681 37 20 271-1, Telefax +49 681 37 20 271-0
Email: info@svh-verlag.de

Approved by: München, TU München, Diss., 2012

Herstellung in Deutschland (siehe letzte Seite)
ISBN: 978-3-8381-3348-5

Imprint (only for USA, GB)
Bibliographic information published by the Deutsche Nationalbibliothek: The Deutsche Nationalbibliothek lists this publication in the Deutsche Nationalbibliografie; detailed bibliographic data are available in the Internet at http://dnb.d-nb.de.
Any brand names and product names mentioned in this book are subject to trademark, brand or patent protection and are trademarks or registered trademarks of their respective holders. The use of brand names, product names, common names, trade names, product descriptions etc. even without a particular marking in this works is in no way to be construed to mean that such names may be regarded as unrestricted in respect of trademark and brand protection legislation and could thus be used by anyone.

Cover image: www.ingimage.com

Publisher: Südwestdeutscher Verlag für Hochschulschriften GmbH & Co. KG
Heinrich-Böcking-Str. 6-8, 66121 Saarbrücken, Germany
Phone +49 681 37 20 271-1, Fax +49 681 37 20 271-0
Email: info@svh-verlag.de

Printed in the U.S.A.
Printed in the U.K. by (see last page)
ISBN: 978-3-8381-3348-5

Copyright © 2012 by the author and Südwestdeutscher Verlag für Hochschulschriften GmbH & Co. KG and licensors
All rights reserved. Saarbrücken 2012

Contents

1	**Introduction**	**1**
2	**General Background**	**5**
2.1	High-Harmonic Generation & Isolated Attosecond Pulses	5
2.2	Attosecond Streaking Spectroscopy	10
	2.2.1 General Principle	11
	2.2.2 Streaking at Surfaces	18
3	**Experimental Setup & Details**	**27**
3.1	Generation of waveform-controlled few-cycle NIR Laser Pulses	27
	3.1.1 The Laser System	28
	3.1.2 Carrier-Envelope Phase Stabilization	31
3.2	Apparatus for Attosecond Electron Spectroscopy in Condensed Matter . .	34
	3.2.1 The NIR–XUV Beamline	35
	3.2.2 Selection of Isolated sub-fs XUV Pulses	38
	3.2.3 The Surface Science End Station	40
3.3	Synchrotron Experiments	43
4	**Surface Electron Dynamics probed by Attosecond Streaking**	**47**
4.1	Attosecond Photoemission from Clean Metal Surfaces	48
	4.1.1 The (110) Surface of Tungsten	48
	4.1.2 Impact of Surface Contamination	62
	4.1.3 The (0001) Surface of Magnesium	65
	4.1.4 Discussion	70

		4.1.5	Electron-Plasmon Interactions	75

- 4.2 Effects of Chemisorption: O/W(110) 79
  - 4.2.1 Sample Preparation 80
  - 4.2.2 Streaking Experiments 81
  - 4.2.3 Interpretation 85
- 4.3 Attosecond Dynamics at Metal-Metal-Interfaces 88
  - 4.3.1 Heteroepitaxy of Magnesium on W(110) 89
  - 4.3.2 Layer-Resolved Attosecond Streaking 94
  - 4.3.3 Discussion 101
- 4.4 Attosecond Streaking in Dielectrics: Xe/W(110) 105
  - 4.4.1 Sample Preparation and Characterization 106
  - 4.4.2 Xenon Monolayer 108
  - 4.4.3 Coverage Dependency 113
  - 4.4.4 Streaking of Solid and Gas-Phase Xenon 117

5 Summary, Conclusion & Outlook 123

A Calibration of Time-of-Flight Data 127

B Accuracy of Delay-Extraction Procedures 131

C Time Shifts & Spectral Resolution 137

Bibliography 139

Chapter 1

Introduction

The desire to steer and control the outcome of physical, chemical and biological processes is one of the main driving forces for exploring their dynamics in time-resolved measurements. Traditionally, these experiments utilize short laser pulses for triggering the process while a second delayed pulse probes the subsequent evolution of the system. Pioneered by the work of A.H. Zewail, these pump-probe experiments allowed to observe the formation and rupture of chemical bonds between molecules in real time [1]. Since the time scale of these dynamics is linked to the vibrations and the atomic-scale movements of the nuclei, laser pulses of a few femtoseconds ($1\,\text{fs} = 10^{-15}$ s) in duration are usually sufficient to track the motion of the molecular constituents during a chemical reaction. In the end, however, bond formation proceeds via rearrangement and transfer of electrons. The understanding and control of this atomic-scale *electronic* motion, which unfolds on a much faster time scale, is important in many fields of science and technology. On a fundamental level, the functioning of solar cells, sensors, the catalytic action of surfaces and the working principle of most of today's nano-scale electronic devices relies on the motion of electrons over the length scale of only a few angstrom ($1\,\text{Å} = 10^{-10}$ m), which implies that the relevant dynamics evolve on an attosecond ($1\,\text{asec} = 10^{-18}$ s) time scale [2].

A straightforward extension of the established laser-based pump-probe schemes to resolve electron dynamics in the sub-fs time domain is frustrated by the natural limit for the minimum duration of a light pulse, which is set by the oscillation period of its carrier wave. This limits the duration for pulses in the near-infrared (NIR) and ultraviolet spectral range to a few femtoseconds, and restricts the prospects of breaking the femtosecond barrier for the pulse duration to the shorter wavelength part of the electromagnetic spectrum. Over the last decade, the tremendous progress in laser technology made it possible to produce highly intense laser pulses in the near-infrared spectral region with durations approaching this single-cycle limit, and even to control the electric field within the pulse envelope [3, 4]. When atoms are exposed to such intense light fields, they can be forced to emit high harmonics of the fundamental laser frequency in form of isolated light bursts of extreme-ultraviolet radiation (XUV), with durations significantly smaller than the period of the NIR driver pulses [5, 6, 7, 8]. Further improvements and developments eventually

Chapter 1. Introduction

cumulated in the generation of XUV pulses with durations of only 80 asec [9]. These are the shortest pulses ever produced in the laboratory. They opened the door for time-resolved spectroscopy with unprecedented temporal resolution [2]. In combination with the strong NIR driving field, these extremely short XUV pulses have been successfully employed to study the dynamics of electrons in simple atoms in the gas phase, where they enabled the observation of electron tunneling [10] and the motion of electrons within the outer valence shells of an atom in real time [11]. Compared to these types of experiments, the application of attosecond spectroscopy to condensed matter systems is still in its infancy.

Today, most of our knowledge on the electronic properties of solids is based on the analysis and interpretation of photoemission experiments. In this fundamental process of light-matter interaction an electron is ejected from the solid upon absorption of a photon. Measuring the energy and momentum of the emitted electrons allows mapping the energy bands, the elemental composition and the atomic structure of a solid [12, 13]. In its time-resolved fashion, two-photon photoemission with optical laser pulses is capable of addressing femtosecond phenomena like thermalization of hot electrons, lifetimes of image-potential and surface states, electron-electron scattering rates or charge carrier relaxation (see e.g. [14, 15, 16]). However, many essential processes in solids and at surfaces, e.g. screening [17], heterogeneous charge transfer [18] and the build-up of electronic band structure [19, 20], can proceed within less than a femtosecond. This calls for spectroscopic techniques offering temporal resolution in the attosecond regime. So far, investigations in this time domain were based on resonant photoemission techniques, which can yield information on electron delocalization rates by using well-known lifetimes of core-level vacancies as an internal time reference [21]. The applicability of this approach is intrinsically restricted by the constraint that the core-level lifetime has to match the time constant of the process to be investigated. Furthermore, this technique only yields the rate of the process but not its complete evolution over time. It is therefore highly desirable to extend the attosecond spectroscopic tools, which have proven to be successful in many gas-phase experiments, to solid-state systems.

The first step in this direction was taken by Cavalieri *et al.* in 2007, where photoelectron emission from a tungsten surface was initiated with attosecond XUV pulses in the presence of the strong electric field of a coincident NIR laser pulse [22]. Upon varying the relative delay between the two pulses, characteristic modulations in the kinetic energy spectra of the photoelectrons could be detected. From the analysis of a sequence of such "streaked" photoelectron spectra, a time delay of only ∼100 asec between electrons released from the core-level and conduction band states of the solid could be inferred. This constitutes the first time-domain measurement in condensed matter with a resolution well below one femtosecond. Despite the existence of several theoretical studies based on the result of this proof-of-principle experiment, it is presently not unequivocally clarified what kind of information is encoded in time delays measured by this attosecond streaking technique, and how it can be related to the electron dynamics taking place within the solid. The main objective of this thesis is therefore to explore the potential of attosecond streaking spectroscopy to investigate surface electron dynamics, and to elucidate the main mechanisms behind the time delays observed in photoemission from solids. To this end, attosecond photoemission experiments on well-defined model systems, including metal single crystals and surfaces covered by adsorbate layers with different physical and chemical properties have been performed. As a prerequisite for this systematic study, a new experimental setup was developed that enables attosecond photoemission experiments under ultra-high vacuum conditions.

This thesis is therefore organized as follows: Chapter 2 describes the basic principles of the generation of attosecond XUV pulses by high-harmonic generation in gases and introduces the concept of attosecond streaking spectroscopy as a technique to resolve ultrafast electron dynamics. Chapter 3 gives specifics on the experimental setup, with emphasis on the ultra-high vacuum apparatus which was developed and set up in the framework of this thesis. The results obtained by attosecond streaking spectroscopy from various well-defined solid-state systems are presented and discussed in Chapter 4. Finally, a summary and outlook conclude this work.

Chapter 1. Introduction

Chapter 2

General Background

This chapter describes the generation of isolated attosecond pulses via high-harmonic up-conversion of laser frequencies in rare-gases and introduces the concept of attosecond streaking spectroscopy. The only experimental result obtained by this method from a condensed matter system so far is also briefly reviewed, along with its current state of interpretation. A detailed introduction to the generation and mathematical description of ultrashort laser pulses [23, 24], as well as to the theory of solid-state photoemission [12, 13] and other basic surface science techniques [25, 26] employed in this thesis, can be found in standard textbooks.

2.1 High-Harmonic Generation & Isolated Attosecond Pulses

In any time-resolved measurement, a reference is needed for sampling that varies on a time scale similar to the dynamical process under investigation. In conventional time-resolved two-photon photoemission, this quantity is the envelope of the laser pulse which limits the achievable resolution to a few femtoseconds. Advancing the photoemission technique to gain access to attosecond phenomena therefore requires flashes of light that last shorter than a femtosecond. Despite the progress made at large-scale Free-Electron-Laser (FEL) facilities, such extremely short light bursts can currently only be produced in the extreme-ultraviolet (XUV) spectral range by generating very high harmonics of optical frequencies with high-intensity laser pulses.

High-harmonic radiation was originally observed in 1987, when intense linearly polarized laser pulses were focused onto atoms in the gas phase [27]. The resulting spectrum is a line series comprising discrete harmonics separated by twice the fundamental frequency ω_L of the driving laser. Only odd harmonics are generated in this process due to the inversion symmetry of the gaseous medium. The generic structure of such a harmonic spectrum is depicted in Fig. 2.1 (a). Whereas the intensity of the lower-order harmonics decreases

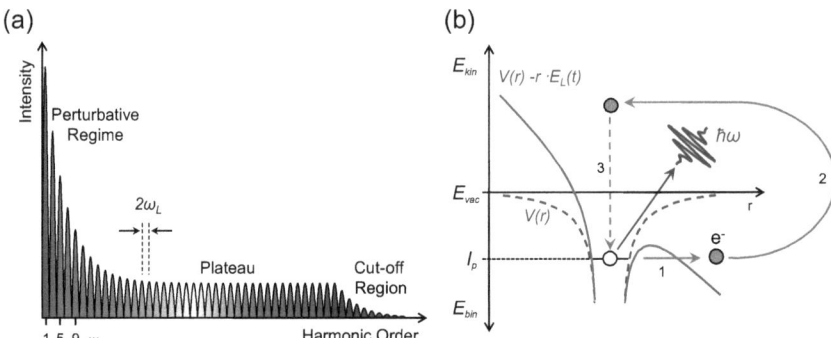

Figure 2.1: Principle of high harmonic generation (HHG) with intense laser pulses. (a) Schematic structure of the photon energy spectrum produced by HHG in atomic gases. (b) Semi-classical model for the HHG process consisting of: (1) optical field ionization, (2) acceleration of the freed electron in the laser field and (3) radiative recombination with the parent ion.

exponentially, the harmonics of higher order are produced with almost constant efficiency giving rise to a plateau extending up to the so-called cut-off region, where the intensity falls off sharply again. The occurrence of a harmonic plateau cannot be explained within standard perturbation theory of nonlinear light-matter interactions, which predicts the efficiency of a nonlinear processes to decrease rapidly with its order. This failure of a perturbative description is related to the enormous strength of the applied electric field. For intensities above 10^{13} W/cm^2, the laser field strength becomes comparable to the binding potential of electrons in the outermost valence shell of an atom and can therefore not be considered as a small perturbation anymore.

The basic mechanism of high-harmonic generation (HHG) can rather be understood from a semi-classical model proposed by Corkum [28] in which the process is divided into three sequential steps (see Fig 2.1 (b)). In the first step, a weakly bound valence electron is detached from the atom by tunnel ionization. This purely quantum-mechanical phenomenon is triggered most efficiently around the local extrema of the optical field oscillation, where the instantaneous electric field is strong enough to substantially deform the atomic Coulomb potential. Once freed from the atomic potential, the electron will be accelerated away from the nucleus like a classical particle in the laser field. A quarter of an optical cycle after ionization, as the electric field reverses direction, the electron will be decelerated again and is finally driven back to its parent ion with a small probability to recombine. The energy gained by the electron during its excursion in the laser field (plus the ionization potential of the atom) will then be released in the form of high energy photons as the atom relaxes to the ground state.

The energy accumulated by an electron when returning to the ion is precisely linked to the

High-Harmonic Generation & Isolated Attosecond Pulses

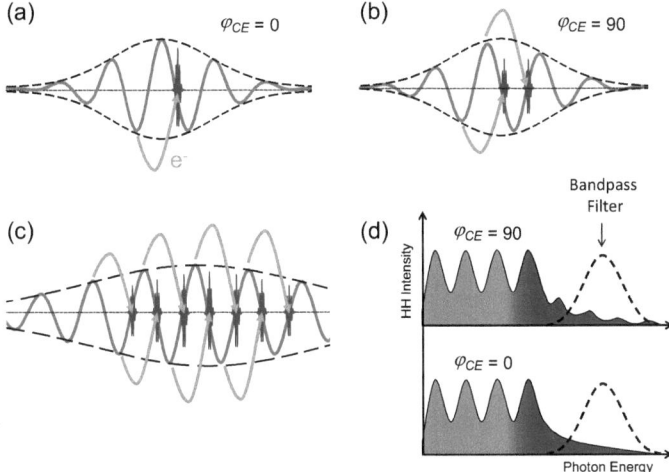

Figure 2.2: Impact of duration and waveform of the driving laser pulse on the temporal structure of the high-energy photons (depicted as violet bursts) emitted during the high-harmonic generation process for (a) a few-cycle pulse with carrier-envelope phase $\varphi_{CE} = 0°$, (b) $\varphi_{CE} = 90°$ and (c) for a multi-cycle driving pulse. The green arrows illustrate the electron trajectories with the highest re-collision energies. (d) The cut-off region of the resultant HH spectrum for $\varphi_{CE} = 90°$ is modulated, but turns into a continuum for $\varphi_{CE} = 0°$. The latter is a signature of a single attosecond pulse in the time domain which can be separated from the residual harmonic radiation by means of a suitable spectral bandpass filter (dashed line). Similarly, the filtered cut-off radiation produced by few-cycle pulse with $\varphi_{CE} = 90°$ will give rise to a double pulse. For longer laser pulses, the cut-off region remains modulated for all settings of φ_{CE} producing a train of attosecond pulses.

instant of ionization within the wave cycle. When the laser pulse comprises many optical cycles, all three steps are repeated periodically for each half-cycle of the driving field. Consequently, the overall emerging harmonic field will be formed by interference of the radiation generated in consecutive half-cycles, which is the origin of the plateau harmonic structure. Because of the strong temporal confinement of the HHG process to less than a wave cycle, these photons are emitted in short bursts which can be significantly shorter in duration than the driving laser pulse. In the time domain, this corresponds to a train of sub-fs pulses separated by half the period of the fundamental laser field [5].

A more detailed analysis of the classical electron trajectories launched during HHG reveals that the maximum energy is transferred from the laser field to the electrons when they are ionized $\omega_L t = 17°$ after the highest field oscillation [28]. The energy released in the corresponding recombination processes defines the high-energy cut-off of the emitted harmonic radiation according to:

$$E_{cut-off} = \hbar\omega_{max} = I_p + 3.17\, U_p. \tag{2.1}$$

High-Harmonic Generation & Isolated Attosecond Pulses

Here I_p is the ionization potential of the atom and U_p denotes the so-called pondermotive potential, i.e. the cycle-averaged kinetic energy acquired by a free electron in the laser field:

$$U_p = \frac{e^2 E_0^2}{4m\omega_L^2}, \tag{2.2}$$

with E_0 being the amplitude the electric field $E_L(t)$ associated with the laser pulse:

$$E_L(t) = E_0(t)\cos(\omega_L t + \varphi_{CE}) = E_0 f(t)\cos(\omega_L t + \varphi_{CE}). \tag{2.3}$$

The phase term φ_{CE} represents the so-called carrier envelope phase (CEP) which describes the timing of the pulse envelope function $f(t)$ with respect to the cosinusoidal carrier wave.

A fully quantum-mechanical treatment of the HHG process supports the intuitive interpretation provided by this semi-classical model, but yields an additional prefactor of ~ 1.3 for the ionization potential in Eq. 2.1 [29]. Common laser sources used for HHG operate at near-infrared (NIR) wavelengths centered near $\lambda_L = 750$ nm. For typical intensities of $5 \cdot 10^{14}$ W/cm^2, the pondermotive potential becomes $U_p \approx 30$ eV. With neon as target gas ($I_p = 21.6$ eV) Eq. 2.1 implies a HH cut-off energy of $\hbar\omega_{max} \approx 116$ eV, which is far in the XUV spectral range. By using helium as generating medium ($I_p = 24.6$ eV) and applying even higher laser intensities, harmonic radiation extending into the water window [30] and even above 1 keV [31] could be produced. At present, however, the achievable flux in this photon energy range is insufficient for spectroscopic applications.

While the carrier-envelope phase φ_{CE} is of almost no relevance for long pulses, it significantly affects the evolution of the electric field within a short few-cycle laser pulse. This is illustrated in Fig. 2.2 (a) and (b) for the two extreme cases $\varphi_{CE} = 0°$ ("cosine" waveform) and $\varphi_{CE} = 90°$ ("sine" waveform), respectively. This sensitivity has far-reaching consequences for the temporal structure of the cut-off harmonic emission generated with these pulses. Because a "sine" waveform exhibits two field extrema of equal strength, two different electron trajectories will contribute to the cut-off radiation. For a "cosine" waveform on the other hand, the highest-energy photons stem only from a single electron re-collision event. In this case, the generation of the cut-off radiation is completely localized in time which corresponds to a continuum in the spectral domain (see Fig. 2.2 (d)) [4, 8]. An experimental example of this phenomenon is shown in Fig. 3.6 on page 37 of this thesis. By selecting a certain bandwidth of this continuum, a single sub-fs pulse can be filtered from the residual HH radiation. Usually, multilayer mirrors consisting of periodic or aperiodic stacks of alternating layers from different materials are used for this purpose. In principle, a suitable choice of both the layer materials, periods and thicknesses allows not only for engineering the center and bandwidth of the reflectivity response, but even for controlling the spectral phase of the reflected XUV pulses [32, 33]. Following this approach, pulse durations as short as 80 asec have been demonstrated [9].

Isolated attosecond pulses are much easier to implement in pump-probe-type experiments than attosecond pulse trains. Filtering the cut-off radiation driven by a sine-shaped laser waveform results in a pair of identical attosecond pulses which is also unfavorable for many applications, especially for those discussed in this thesis. Moreover, any deviation from the "cosine" waveform will give rise to a satellite burst accompanying the main attosecond pulse, with the intensity of the satellite depending sensitively on the CEP [34]. In this respect it has to be mentioned that for driver pulses with durations very close to the single cycle limit the range of CEP values for which single attosecond pulses can be filtered with negligible satellite content is more extended. This is a consequence of the large contrast in amplitudes for adjacent field extrema in these waveforms [9]. In general, pulses delivered by a laser system do not have a constant waveform because the CEP varies from shot to shot. For a reproducible and reliable generation of single attosecond pulses, the control and stabilization of the CEP is therefore of uttermost importance. Experimental schemes to control the waveform of laser pulses will be discussed in Section 3.1.2.

Due to the generation process, the higher harmonics also inherit the spatial and temporal coherence properties from the fundamental laser field. The harmonic cut-off emission is therefore radiated with a very small divergence angle θ_n that roughly scales with the harmonic order n according to $\theta_n \approx \theta_L/\sqrt{n}$, where θ_L is the divergence angle of the driving laser beam [35, 36]. This property is crucial for experiments, where a spatial separation of the generated XUV radiation from the fundamental light is often required.

Most of the above considerations concerning HHG applied to the response of a single atom. In practice, however, the HH radiation is generated from a macroscopic ensemble of atoms confined in a gas target to achieve sufficiently high photon flux. In order to build up coherently over an extended propagation distance, the harmonics and the fundamental laser light have to travel with the same phase velocity (phase matching). Only then, harmonics generated from different atoms in the interaction region will add constructively to the final harmonic field. Notably the dispersion induced by the free electrons produced during the HHG process and the geometric phase shift introduced by focusing of the laser beam (Gouy phase shift [37]) contribute to a dephasing between the fundamental and its higher harmonics. Even though these effects can be partly accounted for by proper adjustment of the interaction length, the gas density, and the focusing conditions, the ultimate conversion efficiency will be limited by re-absorption of the XUV photons by the surrounding gas atoms to $\sim 10^{-5} - 10^{-6}$ [3, 38].

The strategy for generating isolated attosecond pulses outlined above heavily relies on intense, few-cycle NIR laser pulses, the generation of which is rather challenging. Different schemes have therefore been developed to obtain single attosecond pulses from longer driving pulses. Promising results have been obtained by so-called polarization gating techniques that make use of the strong dependence of the HHG process on the ellipticity of the driving laser field [39]. This sensitivity can be readily understood from the intuitive model for HHG, since electrons tunnel-ionized by a non-linearly polarized light field will never return to their parent ion. Thus, employing pulses with a time-varying polarization state, it is possible to confine HHG to a time interval that can be shorter than half an optical cycle of the fundamental field. The shortest XUV pulses generated in this way have a duration of 130 asec and are centered at ~ 36 eV [40]. The requirements concerning the laser pulse duration are even more relaxed when this gating method is combined with two-color fields, e.g. the superposition of the fundamental laser field with its second harmonic. Driven by such a laser field, the inversion symmetry in the HHG process is broken which allows the generation of both even and odd harmonics. In the time domain, this translates into a temporal separation equal to the full period of the fundamental laser between consecutive pulses in the pulse train, which puts less stringent demands on the width of the temporal gate. With this so-called "double optical gating" method it seems to be possible to generate isolated XUV pulses with durations of ∼130 asec directly from 9 fs long NIR driver pulses [41].

Nevertheless, the use of sub-4 fs driving laser pulses still offers several important advantages, such as the potential to reach higher cut-off energies and better conversion efficiencies due to favorable phase matching conditions [42, 3]. Both aspects are crucial for the feasibility of most of the attosecond experiments presented in Chapter 4.

2.2 Attosecond Streaking Spectroscopy

Although isolated sub-fs XUV pulses are becoming more and more available today, the associated low photon flux still impedes their application in conventional pump-probe schemes, where an sub-fs XUV pulse (resonantly) excites the electronic system of the sample, whose temporal evolution is then probed by absorption, reflection or photo-ionization initiated by a second delayed attosecond pulse. On the other hand, these attosecond pulses are inherently synchronized to the electric field of the NIR driver pulses used for their generation via high-harmonic up-conversion (see previous section). Therefore, most of the attosecond experiments demonstrated so far draw on the common principle to substitute either the pump [11] or the probe pulse [43, 10, 22, 44] with these controlled laser fields. Especially in experiments based on attosecond single-photon ionization, the necessary time resolution can currently only be achieved by "dressing" the XUV-induced photoemission with a strong NIR field. In these measurements, the energies of the electrons photo-emitted by the XUV pulse are further modulated in the presence of the laser field, leading to characteristic shifts and distortions of the photoelectron spectrum as a function of the relative delay between the NIR and XUV pulses [45]. This method is commonly referred to as "attosecond streaking" in the limit of XUV pulse durations smaller than the period of the NIR field oscillation ($T_L \approx 2.6$ fs). It was originally conceived and employed for the temporal characterization of the attosecond pulses themselves [45, 46, 6, 7, 34].

The basic concept of attosecond streaking will be outlined in the first part of this section, since all the time-resolved experiments presented in this thesis strictly rely on this technique. The second part gives a short overview of the current state of research in the field of attosecond streaking in solids. Some general particularities pertaining to streaking spectroscopy in condensed matter systems are also addressed.

2.2.1 General Principle

In attosecond streaking experiments, a short XUV pulse generates an electron wave packet by photo-ionizing atoms in the presence of a strong NIR laser field. Insights into the evolution of these electrons can be already gained from an intuitive semi-classical two-step description of the process. First, an XUV pulse with a central photon energy $\hbar\omega_x$ liberates an electron from an energy level with binding energy E_{bin}. A possible influence of the laser field on the initial state is neglected (strong-field approximation). The electron appears in the continuum with an initial momentum $p_i = \sqrt{2m_e(\hbar\omega_x - E_{bin})}$ and is instantaneously accelerated like a classical particle in the oscillating electric field $E_L(t)$ of the laser pulse, which is assumed to be polarized along the z-direction. If the release of the electron into the laser field occurs at the instant of time τ, its momentum component parallel to the laser polarization will be changed by $\Delta p_z(\tau)$ after the laser pulse has left the interaction region. This momentum shift can be calculated by integrating the classical equation of motion for the electron:

$$m_e \ddot{z} = -eE_L(t) = -eE_0(t)\cos(\omega_L t + \varphi_{CE}) \tag{2.4}$$

$$\Rightarrow \Delta p_z(\tau) = m_e \dot{z} = -e \int_\tau^\infty E_0(t)\cos(\omega_L t + \varphi_{CE})\, dt = -eA_L(\tau). \tag{2.5}$$

Here, $A_L(t)$ denotes the vector potential which is related to the electric field by $A_L(t) = \int_t^\infty E_L(t')\, dt'$ (Coulomb gauge). The finite duration of the envelope $E_0(t)$ ensures that only electrons which are set free during the laser pulse will experience a net change of their momentum. Within the adiabatic approximation, i.e. $dE_0/dt \ll E_0\omega_L$, Eq. 2.5 can be further evaluated to:

$$\Delta p_z(\tau) = \frac{eE_0(\tau)}{\omega_L}\sin(\omega_L \tau + \varphi_{CE}) = \sqrt{4U_p(\tau)m_e}\sin(\omega_L \tau + \varphi_{CE}). \tag{2.6}$$

The final momentum p_f of the electron can be obtained from the trigonometric relation $p_f^2 = p_i^2 + 2p_f\Delta p_z \cos\theta - \Delta p_z^2$, where θ is the angle between the laser polarization and the final momentum of the detected electron (see Fig 2.3 (b)). The corresponding final kinetic energy $E_{kin,f}$ is approximately given by [45]:

$$E_{kin,f} = \frac{p_f^2}{2m_e} \approx E_{kin,i} + 2U_p(\tau)\cos 2\theta \sin^2(\omega_L \tau + \varphi_{CE}) \tag{2.7}$$

$$+ \sqrt{8U_p(\tau)E_{kin,i}}\cos\theta\sin(\omega_L \tau + \varphi_{CE}). \tag{2.8}$$

The observed modulation of the final kinetic energy does therefore not only depend on the release time τ, but also on the geometry of the experiment. This is illustrated for an isotropic distribution of initial photoelectron momenta in Fig. 2.3 (b)-(e). When the electrons are detected in the direction perpendicular to the laser polarization ($\theta = 90°$), the measured electron spectra will be asymmetrically broadened and shifted to lower kinetic energies with twice the frequency of the laser field ($2/T_L$) [6, 7]. This purely geometrical broadening is obviously proportional to the acceptance angle of the (angle-integrating)

Attosecond Streaking Spectroscopy

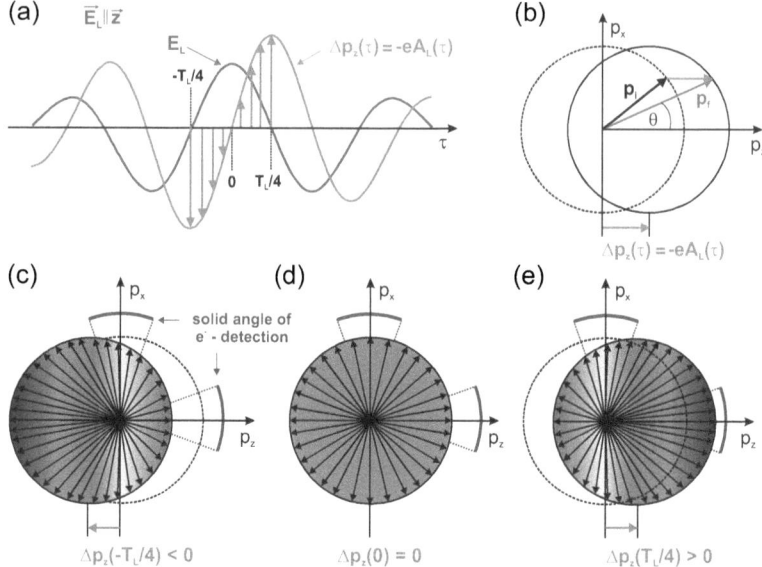

Figure 2.3: Principle of photoelectron streaking with a linearly polarized laser pulse. (a) Electrons released into the NIR field $E_L(t)$ at the instant of time τ suffer a momentum shift Δp_z which is defined by the instantaneous value of the vector potential $A_L(\tau)$. (b)-(e) Illustration of the streaking effect on an isotropic distribution of initial electron momenta. Depending on the angle of electron detection θ, the laser-induced distortion of the momentum distribution will be predominately perceived as a periodic broadening ($\theta = 90°$), or as an alternating energy shift ($\theta = 0°$) in the measured electron energy spectra.

electron analyzer. In contrast, if only electrons that are emitted predominantly along the laser polarization ($\theta = 0°$) are collected, the corresponding photoelectron spectra will be periodically up- and down-shifted in energy without substantial broadening. This parallel detection scheme was adopted for all the measurements presented in this thesis. For this geometry, the modulation of the final electron kinetic energy as a function of τ is given by:

$$\Delta E_{kin}(\tau) = E_{kin,f} - E_{kin,i} \approx \sqrt{8\,U_p(\tau)E_{kin,i}}\,\sin(\omega_L \tau + \varphi_{CE}), \qquad (2.9)$$

provided that $E_{kin,i} = \hbar\omega_x - E_{bin} \gg U_p$. This condition is easily fulfilled in all the streaking experiments that will be presented in Chapter 4. For the highest NIR intensities of $\sim 10^{12}$ W/cm^2 employed in these measurements, U_p amounts only to ~ 50 meV ($\lambda_L = 750$ nm) which is negligible compared to the electron kinetic energies of interest (50 – 130 eV).

Equations 2.5 and 2.9 imply that for parallel detection, the vector potential A_L of the linearly polarized laser pulse can be directly sampled upon varying the release time τ of the photoelectrons into its electric field. This was experimentally confirmed in 2004,

Attosecond Streaking Spectroscopy

where "streaked" photoelectron spectra originating from gas-phase neon were recorded as a function of the relative delay between the NIR and the sub-fs XUV pulses [34, 47]. Since the momentum shift imparted on the electrons also depends on the carrier-envelope phase φ_{CE}, such streaking experiments require active stabilization of the sampled NIR waveform, which in turn is the prerequisite for generating the isolated attosecond XUV pulses needed in these experiments anyhow.

The formalism used above treated the electron as a classical, point-like particle. In reality, however, the launched photoelectrons will form a wave packet whose properties are, in the simplest case, governed by the ionizing attosecond pulse that will imprint its duration and time-frequency-dependence (chirp) onto the released photoelectron distribution. This finite temporal duration of the generated electron wave packet manifests itself in a symmetric broadening of the photoelectron spectra whenever the release of the electrons coincides with a zero-crossings of the NIR vector potential (i.e. local extrema of the electric field), because in this situation electrons within the release distribution can acquire momentum shifts in opposite directions (see Fig. 2.4 (a)). As opposed to this, a chirp in the electron wave packet will lead to a broadening *and* narrowing of the photoelectron spectra, when compared at consecutive zero-crossings of the vector potential. This is illustrated in Fig. 2.4 (b) for a wave packet carrying a negative chirp[1]. When the electron wave packet is probed at zero-crossings exhibiting a negative slope of the vector potential, the faster electrons in the leading edge will be further accelerated by the laser field, whereas the slower electrons in the trailing edge will be decelerated which results in a broadening of the streaked electron energy spectrum. Conversely, a narrowing of the photoelectron spectrum is observed when the release of the wave packet is synchronized with a zero-crossing of the vector potential where the slope is positive. As long as the photoemission process can be considered instantaneous, a streaking experiment can be interpreted as a cross-correlation between the attosecond XUV pulse and the NIR laser pulse. It has been demonstrated that the analysis of streaking measurements with FROG[2]-type algorithms can provide almost complete information about the temporal intensity profile and the spectral phase of both the XUV pulses and the NIR electric field [48, 49, 50].

In quantum mechanics, the temporal evolution of an electron wave packet $|\psi(t)\rangle$ in a laser field with vector potential $\mathbf{A}_L(t)$ is governed by the time-dependent Schrödinger equation (TDSE)[3]:

$$i\frac{\partial}{\partial t}|\psi(t)\rangle = \hat{H}(t)|\psi(t)\rangle. \qquad (2.10)$$

In dipole approximation, the Hamiltonian $\hat{H}$ can be written as:

$$\hat{H}(t) = \frac{1}{2}(\hat{\mathbf{p}} + \mathbf{A}_L(t))^2 + V_{eff}(\mathbf{r}) + \mathbf{E}_X(t) \cdot \mathbf{r}, \qquad (2.11)$$

where only a single electron moving in an effective potential $V_{eff}(\mathbf{r})$ formed by the ion and the remaining bound electrons is considered (single active electron approximation). As

---
[1] In a negatively chirped wave packet, the instantaneous frequency (energy) decreases with time.
[2] Frequency-Resolved Optical Gating
[3] Atomic units will be used for the remainder of this section.

Attosecond Streaking Spectroscopy

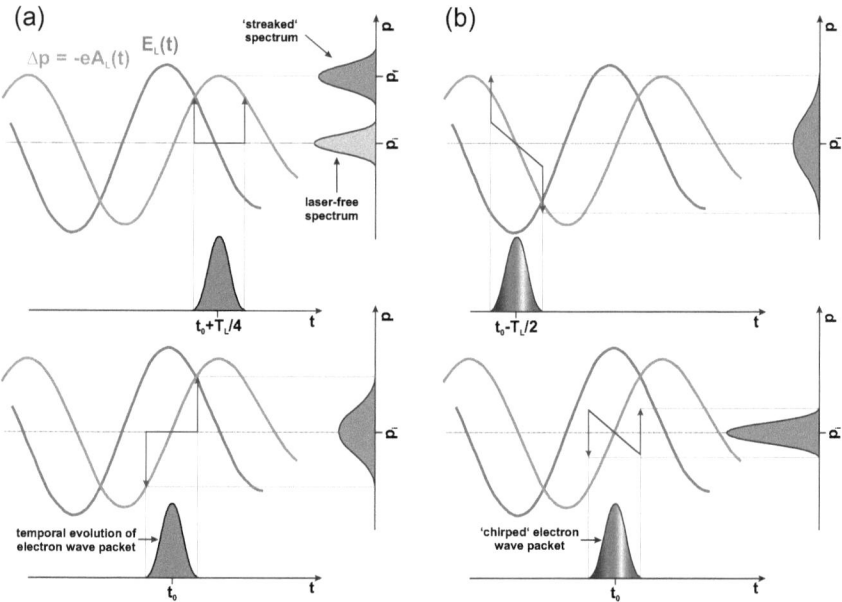

Figure 2.4: Manifestation of the electron wave packet's chirp and duration in attosecond streaking experiments. In the presence of the NIR dressing field $E_L(t)$, the initial time-intensity distribution of the electron wave packet will be mapped onto a corresponding final momentum-intensity distribution (photoelectron energy spectrum) which is observed parallel to the laser polarization. (a) For electron wave packets with constant time-frequency evolution ("unchirped"), the finite wave packet duration gives rise to a broadening of the final momentum distribution whenever the release of the electrons into the laser field coincides with a zero-crossing of the NIR vector potential $A_L(t)$. (b) When the electron wave packet carries a negative (positive) linear chirp, the final momentum distribution will be narrowed (broadened) at zero-crossings with a positive slope of $A_L(t)$, and broadened (narrowed) when the emission is synchronized with zero-crossings featuring a negative slope.

usual, $\hat{p} = -i\nabla$ represents the canonical momentum operator. The XUV pulse responsible for ionizing the atom is described by $\mathbf{E}_X(t) = E_X(t)e^{-i\omega_x t}\,\mathbf{n}_X$, where $E_X(t)$ is the envelope of the pulse, ω_X the central frequency of the radiation and $\mathbf{n}_X$ defines the polarization of the electric field. The transition amplitude $a(\mathbf{p}, \tau)$ for populating the final continuum state $|\mathbf{p}\rangle$ with momentum $\mathbf{p}$ can be calculated as a function of the delay τ between the ionizing XUV pulse and the dressing laser field according to [45, 46, 48]:

$$a(\mathbf{p},\tau) = \langle \mathbf{p}|\psi(t)\rangle = -i\int_{-\infty}^{\infty} dt\, E_X(t+\tau)\, M(\mathbf{p}+\mathbf{A}_L(t))\, e^{i\Phi_v(\mathbf{p},t)}\, e^{i(p^2/2)t - i(p_i^2/2)t}. \quad (2.12)$$

Thus, the main effect of the laser field is to impose an additional phase modulation $\Phi_v(\mathbf{p}, t)$

on the XUV-induced wave packet, which is referred to as Volkov phase [51]:

$$\Phi_v(\mathbf{p}, t) = -\int_t^\infty dt' \left(\mathbf{p}\,\mathbf{A}_L(t') + \frac{1}{2}\mathbf{A}_L^2(t') \right). \tag{2.13}$$

Equation 2.12 relates the electric field of the XUV pulse to the generated electron wave packet and therefore constitutes the general theoretical foundation of most techniques developed for attosecond pulse characterization. The dependence of the streaking effect on the observation angle θ is now contained in the dot product $\mathbf{p}\,\mathbf{A}_L$. The initial central momentum p_i is determined by energy conservation: $E_{kin,i} = p_i^2/2m_e = \hbar\omega_x - E_{bin}$.

The response of the atomic system enters solely through the transition dipole matrix element (TDME) $M(\mathbf{p}) = \langle \mathbf{p} | \mathbf{r} \cdot \mathbf{n}_X | \psi(t_0) \rangle$, evaluated between the ground state $|\psi(t_0)\rangle$ and a plane wave $|\mathbf{p}\rangle$ describing a free electron with asymptotic final momentum $\mathbf{p}$. Any influence of the atomic Coulomb potential on the propagation of the electron wave packet is thereby neglected (strong-field approximation). According to Eq. 2.12, both the ionizing attosecond pulse and the TDME will shape the outgoing wave packet. Especially, when the excitation energy is close to a resonance of the atom (e.g. in auto-ionization), the TDME can vary strongly over the spectral bandwidth of the attosecond pulse. In these situations, the amplitude and phase of the released electron wave packet will differ from the corresponding properties of the XUV pulse [52, 53].

When neglecting a possible energy-dependence of the TDME, a simple time-domain electron wave packet $\chi(t)$ may be constructed by:

$$\chi(t) = f(t)\, e^{-i(p_i^2/2)t}, \tag{2.14}$$

where the temporal evolution of both the attosecond electric field and the TDME are absorbed in a single complex-valued function $f(t)$. In the following, the electrons are assumed to be detected along the laser polarization. The streaked photoelectron energy spectrum $P(E, \tau)$ in function of τ is then given by:

$$P(E, \tau) = P(p^2/2, \tau) = |\chi(p, \tau)|^2 = \left| -i \int_{-\infty}^{\infty} dt\, f(t)\, e^{i\Phi_v(p,t)} e^{i(p^2/2)t - i(p_i^2/2)t} \right|^2. \tag{2.15}$$

This expression reduces to a simple Fourier transform when the central momentum approximation $\Phi_v(p, t) \approx \Phi_v(p_i, t)$ is applied. It is instructive to calculate the corresponding streaked electron spectra using Gaussian parameterizations for $f(t)$ and the laser vector potential $A_L(t)$:

$$f(t) = f_0\, e^{-4\ln 2\,(t/\tau_x)^2} e^{i b_x t^2} \tag{2.16}$$

$$A_L(t) = A_0\, e^{-4\ln 2\,(t/\tau_L)^2} \sin(\omega_L t + \varphi_{CE}), \tag{2.17}$$

with τ_x and τ_L being the full width at half maximum (FWHM) duration of the electron wave packet and the NIR laser pulse, respectively. The phase factor b_x defines a linear

15

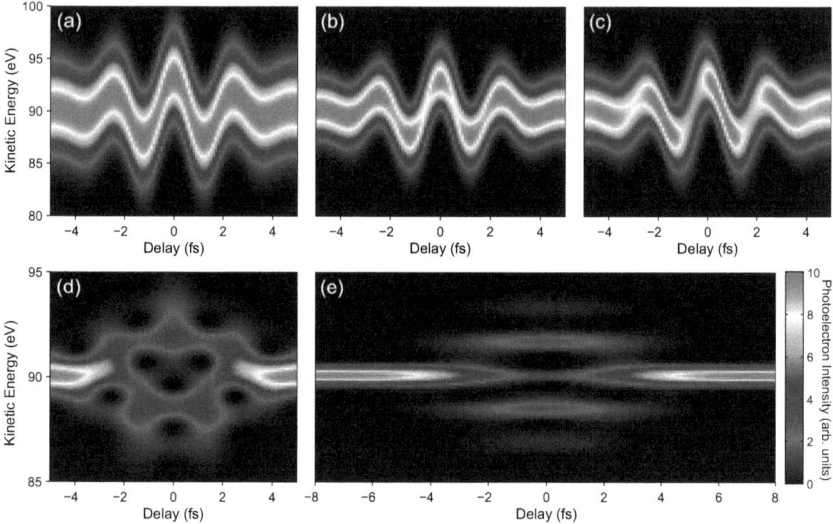

Figure 2.5: False-color images of streaking spectrograms calculated from Eq. 2.15 for a Gaussian electron wave packet with duration τ_x released into the electric field of a 5 fs long NIR pulse with a carrier wavelength $\lambda_L = 750$ nm and an intensity of $2 \cdot 10^{11}$ W/cm^2. Spectrograms are calculated for (a) $\tau_x = 300$ asec, (b) $\tau_x = 450$ asec, (c) $\tau_x = 450$ asec with a negative linear chirp $b_x = -3$ fs^{-2}, (d) $\tau_x = 1.5$ fs and (e) $\tau_x = 3$ fs. When the duration of the wave packet approaches half the period of the NIR field ($T_L/2 = 1.3$ fs), the effect of dressing the photoemission with the laser field develops from streaking into the generation of a series of sidebands, which are spaced in multiples of $\hbar\omega_L = 1.65$ eV from the main photoemission line.

chirp of the wave packet, and the constant A_0 is determined by the intensity of the streaking field.

A compilation of laser-dressed photoelectron spectra $P(E, \tau)$ for different relative delays τ between the attosecond XUV pulse and the laser pulse is referred to as streaking spectrogram. Figure 2.5 shows false-color representations of spectrograms calculated according to Eq. 2.15 for different durations of the electron wave packet with a central kinetic energy $E_{kin,i} = 90$ eV. As long as the wave packet duration is significantly shorter than half the period of the streaking field (panels (a)-(c)), the centroids of the streaked electron spectra follow the evolution of the vector potential of the laser field. This is in agreement with the classical analysis, and is a simple consequence of the one-to-one correspondence between the change in momentum imparted to the emitted electrons and the change of the electric field (see Fig. 2.3 (a)). Furthermore, also the effect of wave packet chirp is reflected in the simulated spectrograms as expected from the simple qualitative discussion in Fig. 2.4. In the false-color images, the characteristic narrowing/broadening of the electron spectra is

Attosecond Streaking Spectroscopy

mainly perceived as the increase/decrease of the electron peak intensity (see Fig. 2.5 (c)).

The spectral appearance of the laser-dressed photoemission changes drastically when the duration of the electron wave packet becomes comparable to, or longer than the half-cycle of the NIR field. An increase in the wave packet duration may either be introduced by a long XUV pulse, or a long-lived intermediate electronic state (e.g. for the emission of Auger electrons). In this regime, the streaking effect gradually disappears with increasing duration of the electron wave packet. Instead, sidebands begin to develop which are separated from the main XUV-induced photoemission line by multiples of the laser photon energy $\hbar\omega_L$ (see Fig. 2.5 (d) and (e)). The formation of these sidebands is the result of quantum interference between different parts of the same electron wave packet that receive identical momentum shift from the streaking field. This quantum mechanical phenomenon cannot be accounted for by the semi-classical model but might be considered as the absorption and stimulated emission of multiple photons by the photoelectron in the laser field, with the momentum conservation being guaranteed by the presence of the ionized atom.

The discussion presented so far illustrates that the modulation of the energy spectrum of the XUV-induced photoelectrons depends sensitively on the instant of time τ at which the corresponding electron wave packet emerges into the laser-dressed continuum. This sensitivity can be used in time-resolved experiments to measure the relative arrival of photoelectrons, which are ejected from different energy levels, in the electric field of the laser pulse. In the streaking regime, such a difference in arrival time $\Delta\tau$ will manifest itself in an offset along the XUV-NIR delay axis between the sampled vector potential waveforms. This is because for any fixed relative delay between the XUV and NIR pulse, an electron released $\Delta\tau$ later into the streaking field will experience a slightly different phase of the NIR field than the remaining electrons, even though the emission of all the electrons is initiated by the same attosecond XUV pulse. In this way, streaking spectroscopy can provide access to dynamics of photo-excited electrons that happens on a time scale much shorter than the duration of employed attosecond pulses. The time resolution in these kinds of measurements is only limited by the fidelity of extracting such shifts from a streaking spectrogram. Thus, it mainly depends on the separation of the energy levels from which the different electrons are emitted, the statistics of the streaked photoelectron spectra and the strength of the applied streaking field.

Experiments along these lines are also conceivable in the sideband regime, since the magnitude of the sidebands still depends on the release time of the electron wave packet relative to the NIR probing field. In this case, however, only a convolution between the envelopes of the wave packet and the NIR pulse can be inferred by tracking the sideband intensity as a function of the XUV-NIR delay. Finding a sub-fs shift between such broad cross-correlation traces is not very reliable thereby limiting the resolution to a few femtoseconds. In contrast, the sub-cycle modulation in streaking spectroscopy holds promise to push the resolving power for relative timing experiments down to the attosecond range. Nevertheless, it has been demonstrated that the evolution of sidebands can provide direct time-domain insight into inner-shell relaxation dynamics in atoms that proceed within ~10 fs [43, 54].

2.2.2 Streaking at Surfaces

In the previous section, laser-dressed photoemission was introduced for an isolated atom, which is appropriate for describing experiments in the gas phase. The extension of this technique to solid surfaces is more challenging, both regarding the experiment and the theoretical modeling of the underlying dynamics. The main complexity is thereby already contained in the single-XUV-photon ionization process itself. Whereas photoelectrons in gas-phase experiments are ejected into simple continuum states, photoemission from solids is more intricate involving the excitation of electrons into energy band states which arise from the interaction of all the remaining electrons in the solid with the periodic potential of the crystal lattice. For a more intuitive discussion of solid-state photoemission, the whole process is often artificially decomposed into three separate steps (three-step model) [55]:

1. The first step is the optical excitation of the electrons inside the solid. The transition probability is determined by the transition dipole matrix element (TDME) which is evaluated between two one-electron bulk Bloch waves associated with the initially occupied band state $E_i(\mathbf{k}_i)$ and the final band state $E_f(\mathbf{k}_f)$. The initial states of localized core electrons exhibit negligible dispersion, i.e. $E_i(\mathbf{k}_i) = E_{bin}$. Besides a non-vanishing TDME, all observed transitions have to obey energy conservation:

$$E_i(\mathbf{k}_i) + \hbar\omega_x = E_f(\mathbf{k}_f). \tag{2.18}$$

In addition, the wave vector of the electron must be conserved:

$$\mathbf{k}_i + \mathbf{G} = \mathbf{k}_f, \tag{2.19}$$

where $\mathbf{G}$ is a reciprocal lattice vector of the crystal. The wave vector of the photon can be neglected in the XUV spectral range.

2. After excitation, the electrons start propagating towards the surface with their group velocity

$$v_g(\mathbf{k}_f) = \frac{1}{\hbar}\frac{\partial E_f(\mathbf{k}_f)}{\partial \mathbf{k}_f}, \tag{2.20}$$

which is governed by the band dispersion in their final states. On their way to the surface, the photoelectrons suffer energy losses due to inelastic electron-phonon and electron-electron scattering, or the excitation of collective modes of the electronic system (e.g. plasmons). The momentum of the photoelectrons can also be altered by elastic collisions. Only electrons that do not lose energy will contribute to the primary electron spectrum with a characteristic photoemission peak whose position is defined by the corresponding initial state. The inelastically scattered electrons

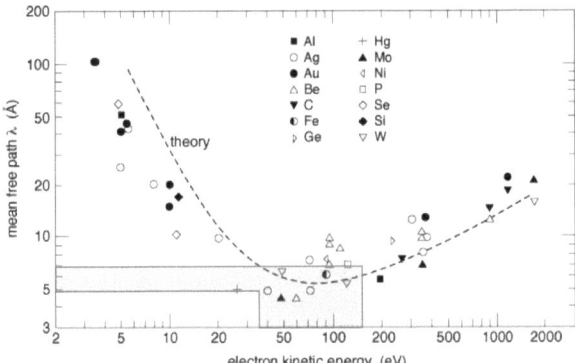

Figure 2.6: Inelastic mean free path (IMFP) of electrons in different materials as a function of their kinetic energy [26]. The electron energy range relevant for the attosecond photoemission experiments presented in this thesis is highlighted.

give rise to a continuous background of secondary electrons at lower kinetic energies. The propagation of excited electrons in solids can be phenomenologically described by the inelastic mean free path λ, which defines the probability $p(z)$ for an electron to lose energy in a scattering event after traveling a distance z through the solid according to $p(z) \propto e^{-z/\lambda}$. Therefore, 63 % of the unscattered electrons escaping from a solid originate, on average, only from a distance λ underneath the surface. For the energy range accessible with XUV photons, this excape depth is less than 8 Å, i.e. only two lattice constants, which renders photoemission from solids extremely surface sensitive in this energy regime. As can be seen in Fig. 2.6, the escape depth is a strong function of the electron energy and is also slightly dependent on the solid-state material.

3. Finally, electrons that arrive at the surface with enough energy to overcome the potential barrier of height V_0 between the solid and the vacuum, will have the possibility to be transmitted and contribute to the final photoemission spectrum. This inner potential V_0, which is roughly given by the sum of the work function ϕ and the width of the occupied part of the conduction band [12], lowers the kinetic energy of the escaping electrons and therefore implies their refraction at the solid-vacuum boundary. In this transmission process, only the wave vector component parallel to the surface is conserved, because of the perturbed translational symmetry normal to the surface.

Although this three-step model is often sufficient to reproduce the main features of a photoemission experiment, it is nevertheless an approximation since it neglects different pathways for the release of a photoelectron that arise from the interference between electron excitation, transport and transmission. In a rigorous quantum-mechanical approach, the whole photoemission event is therefore treated as a single coherent process. Within such a one-step model, the transition matrix element is usually taken between a bulk Bloch state inside the crystal and a so-called "time-reversed LEED[4] state". This final state is constructed by matching propagating bulk Bloch waves inside of the crystal to plane wave states in vacuum that describe the free electrons traveling to the detector [56, 12, 13]. As a result of this constraint to match the wave functions at the interface, only a fraction of the final Bloch band states that are accessible by optical excitation can actually contribute to the emergent photoemission current. The finite escape depth of the electrons λ is usually accounted for by adding an imaginary part to the inner potential, which leads to an exponential damping of the time-reversed LEED state in the interior of the solid. The mathematical formalism needed for a proper description of this one-step approach is omitted here, since a discussion based on the more phenomenological three-step model turned out to be sufficient for interpreting most of the results obtained in this thesis.

From this brief introduction it is already evident that an electron wave packet created in a solid by the absorption of an attosecond XUV pulse can exhibit a more complex time evolution than a corresponding wave packet released from an isolated atom in the gas phase. Especially the electron wave packets launched from the conduction band of a solid comprise a manifold of transitions with different final-state dispersions. Since the final band states disperse with the electron wave vector $\mathbf{k}_f$, the relevant group velocities v_g of the detected electrons will not only depend on the XUV energy $\hbar\omega_x$, but also on the observation geometry defined in the experiment. Moreover, the matrix elements for the individual transitions may vary substantially within the bandwidth $\Delta\hbar\omega_x$ of the exciting attosecond radiation, especially when surface states are involved [57]. Finally, even the electron wave packets released from the localized core states of the solid, which could be thought to resemble closest the attosecond photoemission process in gases, are actually formed by the coherent superpositions of all photoelectrons emitted from the periodically arranged atoms in the lattice of the crystal which may give rise to interference phenomena.

When the XUV-induced photoemission is combined with an intense NIR pulse for time-resolved experiments, a further complication concerning the spatial inhomogeneity of both the intensity and polarization of the laser field arises because of screening and refraction of the incident NIR light at the surface of the solid. In general, the photoelectrons produced by an attosecond pulse will therefore be subject to a spatially varying streaking field while propagating towards the surface. This spatial inhomogeneity of the probing field obviously depends on the optical properties of the solid-state material. On the experimental side, the NIR laser intensities that can be applied to dress the primary photoemission from a solid are generally restricted to much lower values than commonly employed in gas-phase experiments in order to avoid irreversible damaging of the crystal surface by laser-ablation. Furthermore, the NIR pulses impinging on the surface are able

---

[4]Low Energy Electron Diffraction

to generate a significant amount of photoelectrons with comparably high kinetic energies by multi-photon ionization. In solids, this process is favored by the low work function which determines the threshold for multi-photon ionization. Such above-threshold photoemission (ATP) electrons contribute to an unwanted background signal which scales with the intensity of the NIR pulses and tends to obscure the observation of the primary XUV-induced photoemission lines. In contrast, for corresponding experiments in the gas phase, which have up to now almost exclusively been performed on rare gas atoms, the threshold for ATP is determined by the ionization potential of the atoms which exceeds the typical work functions of solids by more than a factor of two and therefore reduces the ATP background considerably.

Laser-assisted photoemission from a solid surface was first observed in 2006 by Miaja-Avila et al., who measured the magnitude of sidebands occurring in the conduction band emission from the Pt(111) surface as a function of the NIR-XUV delay [58]. From the resultant cross-correlation trace, the 10 fs duration of the employed XUV pulses could be inferred but further insights into the dynamics of the photo-generated electrons inside the solid could not be obtained. In 2007, Cavalieri et al. reported the first application of streaking spectroscopy in condensed matter [22]. In this experiment, the streaked photoemission originating from the (110) surface of tungsten could be measured. The scenario of the experiment is schematically depicted in Fig. 2.7. The employed XUV pulses had a duration of $\tau_x \approx 300$ asec and a central energy of $\hbar\omega_x \approx 90$ eV. The streaking spectrogram featured two broad peaks related to the photoemission from the conduction band (CB) and the $4f$ core states of tungsten, which both exhibited the characteristic energy modulation tracing the waveform of the NIR vector potential of the $\sim$5 fs long streaking pulses. Further analysis revealed a shift of $\Delta\tau = 110 \pm 70$ asec between the two parts of the spectrogram. Since both types of electrons are excited simultaneously by the attosecond XUV pulses[5], this time shift could be interpreted as a transport effect associated with the propagation of the photoelectron wave packets towards the surface of the crystal.

Since the electron emission was collected along the surface normal, only the NIR field component parallel to this direction is relevant for the observed streaking effect[6]. In the experiment, the p-polarized NIR and XUV pulses illuminated the surface under $\vartheta = 15°$ gracing incidence (see Fig. 2.7). For this angle of incidence, Fresnel equations predict the normal component of the streaking field inside the W(110) crystal to be reduced to only 6 % of the corresponding field strength in vacuum [22]. According to this estimate, the effect of streaking is negligible before the electrons traverse the metal surface. Consequently, the measured time delay may be linked to a difference in arrival time of the W$4f$ and CB electron wave packet at the surface of the metal. A delayed emssion of the less energetic W$4f$ electrons is then indeed expected, since their mean escape depth is $\sim$1 Å larger than for the CB electrons [22]. If the excited electrons behave like free electrons inside

---

[5]A photon covers a distance of $\lambda \approx 8$ Å in less than 3 asec.
[6]For a perfect conductor, the electric field component parallel to the surface has to vanish anyhow.

Attosecond Streaking Spectroscopy

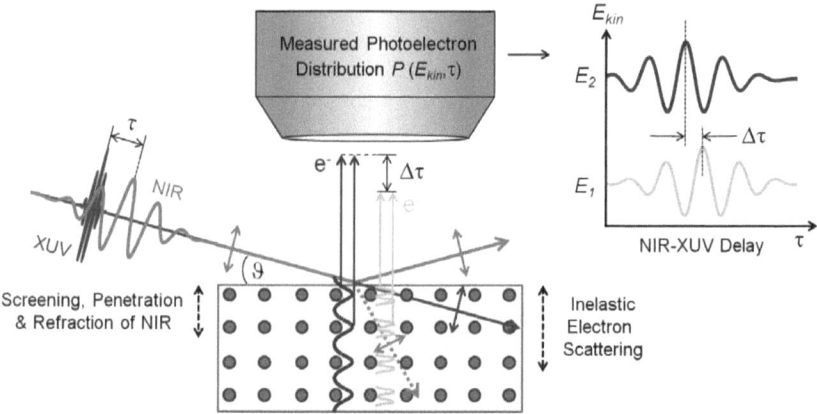

Figure 2.7: Scenario for attosecond streaking at a solid surface. The attosecond XUV pulses photoexcite electrons from different initial states of the solid within a region exceeding the escape depth of the electrons. Pictorial wave functions for localized core electrons and delocalized conduction band electrons are indicated. The generated electrons propagate towards the surface in a (possibly) spatially inhomogeneous NIR streaking field. Any time delay $\Delta\tau$ occurring between photoelectrons released from different energy levels manifests itself in a corresponding offset $\Delta\tau$ between their streaked electron distributions $P(E_{kin}, \tau)$ which are measured as a function of the relative NIR-XUV pump-probe delay τ. In general, $\Delta\tau$ may be attributed to the interplay of NIR field penetration and refraction at the surface, the strength of inelastic electron scattering, differences in the electron velocity and the initial state character of the involved electronic levels.

the metal, their velocity would be given by:

$$v_{free} = \sqrt{\frac{2}{m_e}(E_{kin} + V_0 - \phi)}, \qquad (2.21)$$

with E_{kin} being the measured kinetic energy of the photoelectrons with respect to the Fermi level and m_e the rest mass of the free electron in vacuum. In this case, a relative time delay of only $\Delta\tau \approx 38$ asec would be expected [22]. The authors argued that the remaining discrepancy with the measured time shift is a signature of the Bloch group velocities associated with the motion of the excited photoelectrons inside the solid. From a simple band structure calculation, considering neither matrix element effects nor the matching conditions for the wave functions at the solid-vacuum interface, a distribution of group velocities $v_g(k_{f,\perp})$ for electrons escaping the W(110) crystal along the surface normal was derived. After averaging these group velocities over the spectral bandwidth of the exciting XUV radiation it was found that electrons from the CB emerge twice as fast from the W(110) crystal as the W4f core-level electrons, thereby increasing their relative time delay to $\Delta\tau = 90$ asec in good agreement with the experiment [22].

This explanation in terms of band-structure-dominated electron transport was challenged in a subsequent theoretical study by Kazansky *et al*, where the time-dependent Schrödinger equation (TDSE) was solved for a one-dimensional (1D) model crystal mimicking the energy levels of W(110) [59]. The interaction of the photoelectrons with the 1D crystal lattice was modeled using a pseudo-potential with parameters originally optimized for the (111) surface of copper, whereas the interaction with the remaining photo-hole in the solid was explicitly taken into account only for electrons released from localized states. Moreover, the NIR field strength was assumed to vanish completely in the interior of the crystal and the escape depth for *both* types of electron was fixed to $\lambda \approx 5$ Å. With these assumptions, the authors calculate a time delay of $\Delta\tau = 85$ asec for the photoelectrons emitted from localized energy level compared to those released from the delocalized band states which is in good agreement with the experimental result. However, within this model the time delay arises primarily from the different spatial confinement of the involved initial states (localized core state vs. delocalized conduction band). The influence of final-state effects was found to be only of the order of $\sim$10 asec. The absence of band structure effects was attributed to the short-term interaction of the outgoing photoelectrons with the crystal lattice, caused by the small electron escape depth and the short duration of the formed electron wave packets, which prevents a modification of the photoelectron velocity by the periodic potential according to Eq. 2.20. This can be seen to be in accord with the interpretation of ion-surface collision experiments, where a diminishing influence of the band gap in Cu(111) on the neutralization rate of H$^-$ ions was found for higher collision energies ($\hat{=}$ short interaction time) [19, 20]. In sharp contrast, recent calculations by Krasovskii *et al.* indicate that, even for almost vanishing escape depths of the electrons, band structure effects may give rise to relative time shifts in attosecond photoemission of up to $\sim$150 asec [60].

A rather different explanation for the observed time delay was inferred from quantum-mechanical calculations by Zhang *et al.* [61, 62]. By adapting the one-step approach of stationary photoemission to laser-dressed emission, transition matrix elements were calculated using damped Volkov wave functions as final states. For the description of the initial states of the metal conduction band, the jellium model was applied which entails completely delocalized wave functions. In accord with [59], the W4f states were modeled with localized atomic-like wave functions. As opposed to [22] and [59], the NIR field was assumed to be constant inside the solid, i.e. any attenuation due to screening and refraction was ignored. Again, the escape depth λ was included as an adjustable parameter, and the time shift of $\Delta\tau = 110$ asec could be reproduced for $\lambda = 2.5$ Å [62]. Furthermore, these calculations suggest that only the temporal evolution of the electron wave packet released from the localized core states is affected by the finite escape depth. The time delay with respect to the CB emission was found to be the result of interference between core-level photoelectrons originating from different layers of the solid. Further, this time shift was predicted to increase monotonically with the inelastic mean free path.

Finally, Lemell *et al.* performed classical transport simulations on the electron release from W(110) for the experimental conditions in the streaking measurement reported by

Cavalieri et al. [63]. The electrons are treated as point-like particles with Gaussian-like initial source distributions spatially confined to the lattice sites of the tungsten crystal. Only the contribution of the $6s$ states to the W(110) conduction band was modeled by a jellium-like distribution. In analogy to [22], the change of the streaking field upon crossing the vacuum-tungsten interface was taken into account by solving the Fresnel equations. Elastic and inelastic scattering cross-sections were derived from calculations based on a muffin-tin potential approximation and the dielectric function of tungsten, respectively. Using the same distribution of group velocities as in [22] for the propagation of the electrons, a time delay of $\Delta\tau = 33$ asec was retrieved from the simulated streaking spectrogram. Similar to [22], the origin of this time delay could be traced back to the difference in velocities and mean free paths of the involved electrons. In addition, the authors find an increase of the time shift to $\Delta\tau = 42$ asec when the NIR field strength inside the crystal is set to zero in the simulations. This was attributed to an effective reduction of the electron escape depth due to enhanced scattering of the electrons moving inside the solid in the presence of the deflecting streaking field. However, both time delays agree only moderately with the experimental result.

In summary, all the above outlined theoretical approaches predict a delayed emission of the $4f$ core electrons in W(110) in accordance with the experimental observation. However, the time delay derived from only a single measurement in [22] comes with a comparably large error margin, making it impossible to favor any of the proposed theories. In addition, all these calculations are based on quite different assumptions concerning the role of band structure for the photoelectron propagation, the initial state localization of the conduction band electrons and the properties of NIR field inside the crystal. Whereas it is incontestable that the NIR field penetrates the solid more than 100 Å (skin depth) [61, 63], the decisive question pertinent to streaking spectroscopy is how abruptly refraction and screening occurs within the first atomic layers of a solid, and therefore reduces the relevant field component parallel to the direction of electron detection. Unfortunately, a microscopic description of these phenomena, that happen on atomic length scales, is rather intricate and presently not available [64]. The application of Fresnel equations relies on a macroscopic (complex) index of refraction and might therefore give only asymptotically correct results. On the other hand, the extremely small electron escape depths for XUV excitation limits the observation of electron dynamics to exactly these first few layers of the solid where deviations from the macroscopic Fresnel field might become relevant. Finally, the inelastic mean free path enters in almost all theories as an adjustable parameter which is generally not known with the precision needed for a meaningful quantitative discussion of few-attosecond time shifts.

A systematic disentanglement of these different contributions to the time shift is challenging, also from the experimental point of view, since all the important parameters and properties of the solid are intertwined. Band structure effects may be disentangled by changing the XUV photon energy to excite the photoelectron wave packets to final states whose dispersion leads to different group velocities. However, also the escape depths will change due to their dependence on the electron kinetic energy. This can be avoided by performing streaking experiments with the same XUV energy, but on different crystalline orientations of the sample. Changing the material of the crystal will obviously alter the escape depths, the character of the CB states and the efficiency of refraction/screening of the NIR field, as well as the electronic band structure. The most direct way to study transport effects is to perform attosecond streaking experiments on a surface covered with a well-controlled number of adlayers. None of the theories published so far can account for the full complexity of a real streaking experiment. It is therefore beneficial to study attosecond photoemission from single crystals featuring a simpler electronic structure than tungsten. However, attosecond streaking at surfaces is still far from being a standard spectroscopy and only a few experimental ideas along these lines could be pursued within the scope of this work. First and foremost, as a prerequisite for any sound interpretation, the existence of time shifts in solid-state photoemission has to be verified. Further, the accuracy of these measurements has to be assessed and improved before the validity of the existing theoretical concepts can be tested. Most of the experiments performed in the course of this thesis are devoted to this fundamental but important aspect.

Attosecond Streaking Spectroscopy

Chapter 3

Experimental Setup & Details

Time-resolved experiments of solid surfaces based on the attosecond streaking principle require a sophisticated infrastructure starting with a laser system delivering ultrashort, intense, waveform-controlled laser pulses, a setup for high-harmonic up-conversion of these optical pulses into the extreme-ultra-violet (XUV) spectral range, suitable filtering of this radiation to isolate single sub-fs XUV pulses, and finally an experimental station allowing the preparation and spectroscopic investigation of well-defined, atomically clean solid-state samples under ultra-high vacuum (UHV) conditions. The latter is mandatory considering the high surface sensitivity of photoemission in the XUV range, which has higher requirements on vacuum conditions than implemented in previous state-of-the-art attosecond pump-probe setups, primarily developed for experiments in the gas phase [65, 66, 67]. This chapter therefore provides a short overview of the main components of the infrastructure developed for performing time-resolved experiments on surfaces with attosecond resolution.

3.1 Generation of waveform-controlled few-cycle NIR Laser Pulses

High-harmonic generation in gases, as discussed in Section 2.1, is a highly non-linear phenomenon and therefore requires powerful laser pulses to achieve the necessary intensity level of $\sim 10^{14}\,\text{W/cm}^2$ to efficiently drive the conversion process. Moreover, these driving laser pulses should comprise merely a few optical cycles of the electric field, the evolution of which has to be precisely controlled in order to reliably generate broadband high harmonic cut-off continua supporting isolated attosecond pulses [8, 4]. This section gives a brief introduction to the laser system employed in all the streaking experiments presented in this thesis. It consists of an ultra-broadband oscillator seeding a Ti:Sapphire multi-pass amplifier to raise the pulse energy, and a subsequent hollow-core fiber/chirped mirror stage for final temporal compression (see Fig. 3.1). This system delivers sub-4 fs pulses in the NIR/VIS spectral range at a repetition rate of 3 kHz with pulse energies up to 400 µJ [68]. The carrier-envelope phase (CEP) of the generated pulses can be stabilized by two independent feedback loops based on common interferometric self-referencing schemes. A more detailed description of the laser system, which was already operational at the beginning of this thesis, can be found in [69].

3.1.1 The Laser System

The front-end of the laser system is a Kerr-lens mode-locked Ti:Sapphire oscillator (modified RAINBOW, Femtolasers GmbH) operating at ~70 MHz. It is pumped by a frequency doubled continuous wave (CW) Nd:YVO$_4$ laser at 532 nm (Verdi, Coherent). The intracavity dispersion is controlled by chirped mirrors [70] allowing for pulse durations down to ~6 fs at the output of the oscillator [71, 72]. Despite their short duration, these pulses cannot be used directly for HHG due to their low pulse energy of only ~4 nJ.

To reach the intensities required for efficient HHG without tight focusing, the energy of the pulses emitted from the oscillator has to be raised by at least 5 orders of magnitude. This is accomplished by a technique known as chirped pulse amplification (CPA) [73]. Before seeding a Ti:Sapphire multi-pass amplifier, the oscillator pulses are temporally stretched to ~15 ps upon propagation through a 5 cm block of flint glass (SF57), which imposes a large positive chirp on the pulses. In this way, the peak power can be kept below the damage threshold of the gain medium and other optical components throughout the subsequent amplification process. In the stretcher, the pulses also undergo multiple reflections on chirped mirrors (HODM), specially designed to pre-compensate for higher-order dispersion accumulated by the pulses in the multi-pass amplifier.

The Ti:Sapphire amplifier is pumped with ~20 W of a frequency-doubled diode-pumped Nd:YLF laser at 532 nm (DM30, Photonics Industries). It is decoupled from the oscillator by a Faraday insulator (FI) to eliminate disturbing influences on the oscillator performance due to back-reflections occurring in the amplifier stage. The stretched pulses are amplified in 9 passes through the Ti-Sapphire crystal which is set up at Brewster's angle to minimize reflection losses, and is thermo-electrically cooled to 190 K to reduce detrimental effects like thermal lensing. After the first four passes, the repetition rate is reduced to ~3 kHz by means of a Pockels cell (PS) synchronized with the amplifier pump laser. The energy of the selected pulses is boosted up to ~1.3 mJ in the remaining 5 passes through the amplifier crystal. A spectral filter (GF) with an inverted Gaussian transmission function centered at a wavelength of ~800 nm is installed in the amplifier beam path to counteract spectral narrowing caused by the limited gain-bandwidth of Ti:Sapphire.

After amplification, the positively chirped pulses are recompressed in a hybrid chirped-mirror/prism compressor. Propagation through two prism pairs arranged in Brewster angle configuration provide the necessary negative dispersion. The prism compressor is modified to overcompensate the positive chirp introduced in the stretcher. In the original design, the shortest pulse occurred inside the last prism of the compression stage. The high

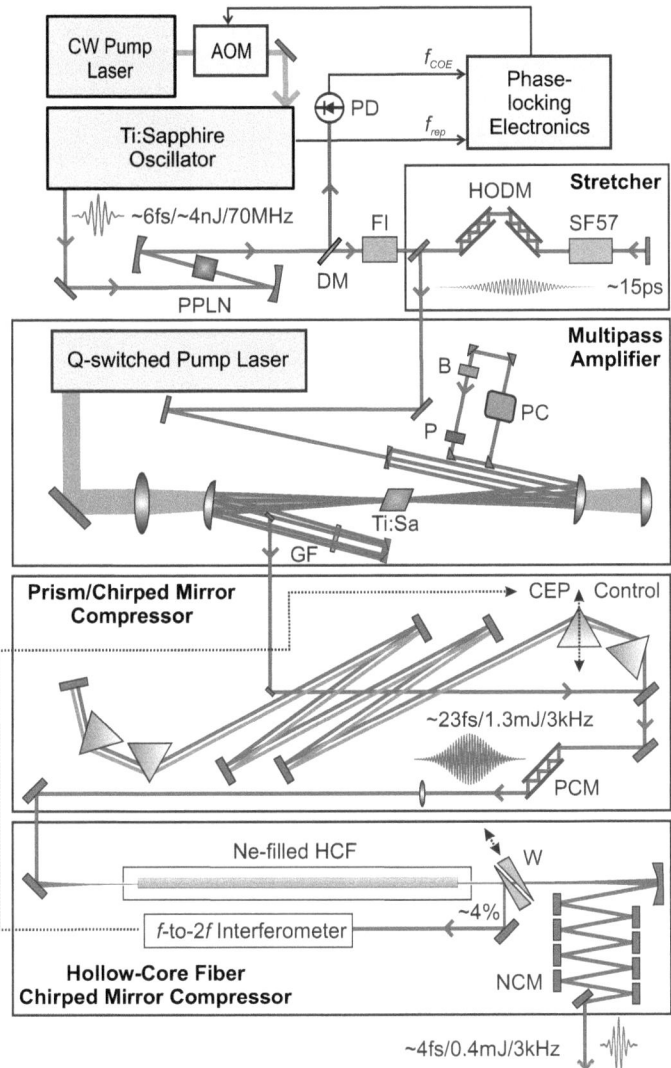

Figure 3.1: Basic layout of the laser system producing waveform-controlled, sub-4 fs pulses in the near-infrared with pulse energies of ∼400 μJ: AOM: acousto-optical modulator, PD: photodiode, DM: dichroic mirror, PPLN: periodically-poled lithium niobate, FI: Faraday isolator, HODM: higher-order dispersion compensation mirrors, SF57: flint glass for pulse stretching, PC: Pockels cell, B: Berek compensator, P: polarizer, Ti:Sa: amplifier crystal, GF: transmission filter, PCM: positively chirped mirrors, HCF: hollow-core fiber, W: fused silica wedges, NCM: negatively chirped mirrors. See text for further explanation.

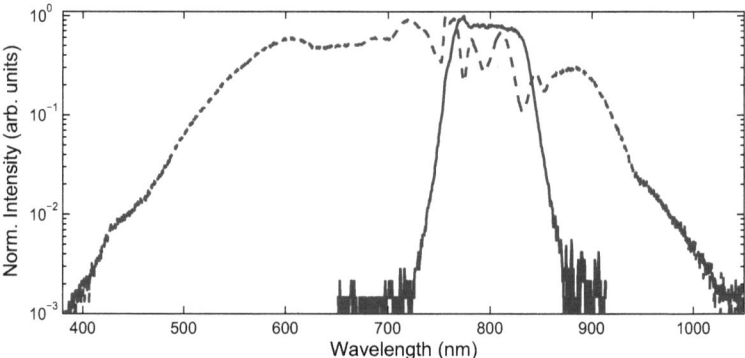

Figure 3.2: Laser spectrum measured after the hybrid chirped mirror/prism compressor (solid line), and after spectral broadening in a hollow-core fiber filled with ∼1.8 mbar neon (dashed line) [68].

peak powers induced self-phase modulation (SPM) within the bulk material of the last prism which in combination with a negative pre-chirp leads to further spectral narrowing [74]. In the current setup, excessive SPM in the prism material is avoided by reducing the peak intensity of the pulses by introducing even more negative chirp. The pulses are then fully compressed to a duration of ∼23 fs upon 14 reflections on positively chirped mirrors (PCM) where SPM is absent.

The duration of the amplified pulses is ultimately limited by the finite gain bandwidth inherent to Ti:Sapphire-based systems. For further temporal compression down to the few-cycle limit, the spectrum of the pulses has to be broadened by generating new frequency components through the non-linear process of self-phase modulation which is based on the intensity-dependent refractive index (optical Kerr effect) [75]. For this purpose, the pulses are focused into a 1 m long hollow-core fiber (HCF) of fused silica (inner diameter: 250 µm) filled with 1.8 mbar neon gas. Typical power transmissions are between 40 − 50 %. A comparison of laser spectra recorded before and after broadening in the HCF is shown in Fig. 3.2.

The efficiency of SPM depends, amongst other parameters, sensitively on the duration of the input pulses [76]. With the shorter pulses resulting from the modified hybrid compressor stage, a NIR/VIS supercontinuum covering a wavelength range of 400 − 1000 nm can be achieved. After removing the positive chirp accumulated during propagation in the HCF by a set of chirped mirrors (NCM) with well-tailored negative dispersion between 550 and 1000 nm, NIR pulses with durations of less than 4 fs (corresponding to only 1.5 optical cycles) and energies of up to ∼400 µJ are available for experiments [68]. A beam stabilization system consisting of two piezo-actuated mirrors installed in front of the HCF further guarantees long-term stability for coupling the laser beam into the fiber. Fine tuning of the pulse compression is provided by a pair of fused silica wedges

(W) inserted into the beam path under Brewster's angle.

3.1.2 Carrier-Envelope Phase Stabilization

Even for ultrashort laser pulses containing only a few optical cycles, it is meaningful to decompose the temporal evolution of the associated linearly polarized electric field $E(t)$ into a carrier and an envelope function according to [3]:

$$E(t) = E_0(t)\cos(\omega_L t + \varphi_{CE}). \qquad (3.1)$$

Here ω_L is the carrier frequency and φ_{CE} denotes the carrier-envelope phase (CEP) as introduced in Section 2.1. Since the carrier wave advances with phase velocity v_p while the envelope $E_0(t)$ moves with the group velocity v_g, the net dispersion experienced by the pulse circulating inside the laser cavity will shift the carrier wave after each round trip with respect to the pulse envelope by:

$$\Delta\varphi_{CE} = \omega_L \left(\frac{L}{v_g} - \frac{L}{v_p}\right), \qquad (3.2)$$

with L being the length of the laser cavity. The corresponding electric field of a pulse train $E_{PT}(t)$ transmitted through the output coupler of a mode-locked laser with a repetition rate f_r can therefore be written as (see Fig. 3.3 (a)):

$$E_{PT}(t) = \sum_{n=-\infty}^{\infty} E_P(t - n\tau)\, e^{i(\omega_L t - n\omega_L \tau + n\Delta\varphi_{CE} + \varphi_0)} \qquad (3.3)$$

where $E_P(t)$ is the envelope of a single pulse, n is an integer, $\tau = f_r^{-1}$ is the cavity round-trip time and φ_0 is a constant phase offset. The Fourier transform of Eq. 3.3 yields:

$$\hat{E}_{PT}(\omega) = \int_{-\infty}^{\infty} E_{PT}(t) e^{-i\omega t} dt \qquad (3.4)$$

$$= \sum_{n=-\infty}^{\infty} e^{i(n\omega_L \tau + n\Delta\varphi_{CE} + \varphi_0)} \int_{-\infty}^{\infty} E_P(t - n\tau) e^{-i(\omega - \omega_L)t} dt \qquad (3.5)$$

$$= e^{i\varphi_0} \hat{E}(\omega - \omega_L) \sum_{n=-\infty}^{\infty} \delta\left(\omega\tau - \Delta\varphi_{CE} - 2\pi n\right) \qquad (3.6)$$

The last equation is obtained by substitution of variables ($t' \to t - n\tau$), defining $\hat{E}_P(\omega) = \int_{-\infty}^{\infty} E_P(t) e^{i\omega t} dt$ and employing the Poisson sum formula [77]. According to Eq. 3.6, the frequency representation of a laser pulse train is a spectrum of equidistant lines separated by f_r as schematically shown in Fig. 3.3 (b). The n-th line of this frequency comb is given by:

$$f_n = \frac{\omega_n}{2\pi} = \frac{\Delta\varphi_{CE}}{2\pi\tau} + \frac{n}{\tau} = f_{CEO} + n f_r, \qquad (3.7)$$

Generation of waveform-controlled few-cycle NIR Laser Pulses

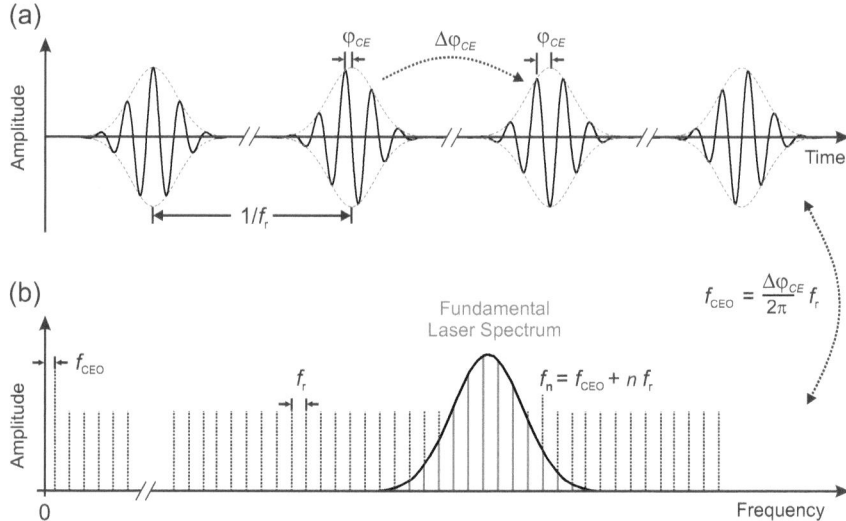

Figure 3.3: Evolution of the carrier-envelope phase φ_{CE} in a pulse train generated by a mode-locked laser. (a) Due to dispersion in the laser cavity, a pulse-to-pulse shift of $\Delta\varphi_{CE}$ occurs between the envelope and the carrier wave. (b) In the frequency domain, this evolution of φ_{CE} prevents the frequency comb lines to be integer multiples of the repetition rate f_r. The offset of the frequency comb f_{COE} equals the recurrence frequency for pulses exhibiting the same carrier-envelope phase. In the absence of active stabilization, the coupling to environmental perturbations causes both f_r and φ_{CE} to evolve in a non-deterministic manner.

where the carrier-envelope offset frequency f_{CEO} describes the rate of change of φ_{CE} within the pulse train. Generally, because of environmental perturbations like thermal effects, air currents and mechanical vibrations, both f_{CEO} and f_r contain noise contributions preventing the frequency comb from being stable.

The measurement and control of optical frequency combs via the radio frequencies f_r and f_{CEO} was a major breakthrough fostering both the fields of attosecond physics and high-precision frequency metrology [78, 79, 80]. Whereas f_r can be readily detected by a fast photodiode inserted into the beam path, f_{CEO} has to be inferred from beat signals produced by interferometric self-referencing techniques [77]. In the so-called f-to-$2f$ method [78], additional frequency combs f_{2n} are produced by second harmonic generation (SHG). Provided the fundamental laser spectrum covers an optical octave, f_{COE} can be observed as heterodyne beat note between the fundamental high-frequency components and the low-frequency part of the SH spectrum (see Fig. 3.4):

$$2f_n - f_{2n} = 2f_{CEO} + 2nf_r - (f_{CEO} + 2nf_r) = f_{CEO}. \tag{3.8}$$

Alternatively, difference frequency generation (DFG) can be used to produce the necessary

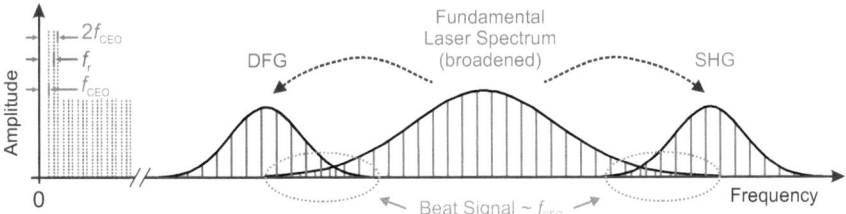

Figure 3.4: Schemes for measuring the carrier-envelope offset frequency f_{CEO} by self-referencing. The interference of the fundamental laser modes with those produced either by second harmonic generation (SHG) or difference frequency generation (DFG) gives rise to a beat note providing information on f_{CEO}. An octave-spanning fundamental laser spectrum is a prerequisite for observing these interference effects.

beat signal in the so-called f-to-0 scheme. DFG between the modes k and m of the laser spectrum results in an intrinsically CEP-stable signal:

$$f_k - f_m = f_{CEO} + kf_r - (f_{CEO} + mf_r) = (m-k)f_r, \quad (3.9)$$

which, however, is too weak for applications like HHG. Nevertheless, information on f_{COE} can be obtained by measuring the beat signal originating from the interference of these DFG modes with the low-frequency components of the fundamental spectrum f_n in the spectral region where $n = m - k$:

$$f_n - (m-k)f_r = f_{CEO} \quad (3.10)$$

The stabilization of the carrier-envelope phase, or more generally the waveform of few-cycle optical pulses, is of vital importance for the generation of isolated attosecond pulses and the feasibility of the streaking experiments which will be presented in Chapter 4. For the laser system employed in this thesis, this control of the CEP is accomplished by two independent stabilization feedback loops for the oscillator and the multi-pass amplifier, respectively. For the oscillator, the f-to-0 scheme is applied. To this end, the output pulses of the oscillator are focused into a periodically-poled lithium-niobate (PPLN) crystal (see Fig. 3.1) where efficient DFG can be initiated due to the high peak power of the ∼6 fs short pulses [81, 82]. Simultaneously, the fundamental laser spectrum is additionally broadened by SPM, leading to a sufficiently strong beat note which can be separated from the fundamental light with a dichroic mirror (DM) and is finally detected by a photodiode.

In view of the low repetition rate of ∼3 kHz supported by the subsequent amplifier, it is sufficient to stabilize the carrier-envelope offset frequency f_{CEO}, which determines the recurrence of a particular CEP in the emitted pulse train, to a fraction of the oscillator repetition rate: $f_{CEO} = f_r/n$. The phase-locking electronics (Menlo Systems) compares the f-to-0 beat signal to an electronically divided reference frequency f_r/n. The resulting error signal is sent into a feedback loop trying to minimize the deviation between both signals by modulating the pump power of the oscillator by means of an acousto-optical modulator (AOM). These intensity modulations change the dispersion inside the laser cavity via the optical Kerr-effect, and therefore influence the evolution of the CEP according to Eq. 3.2. In the current setup, the reference frequency is $f_r/4$, meaning that every fourth pulse in the pulse train seeding the amplifier has the same CEP. The Pockels cell in the multipass amplifier is programmed to selected only a few of these pulses to reduce the repetition rate to $\sim$3 kHz.

Intensity fluctuations and pointing instabilities occurring during the amplification process and in the subsequent compression stages translate into a drift of the CEP when observed at the output of the laser system. Therefore, a second control loop based on the f-to-$2f$ principle is installed right after the HCF, where $\sim$4 % of the output beam is split off and focused into a β-barium borate BBO crystal for SHG. The interference pattern in the region of spectral overlap between the fundamental and its second harmonic, which contains the information on the CEP, is recorded with a fiber spectrometer and Fourier-analyzed. A proportional-integral (PI) controller stabilizes the fringe positions, and therefore the CEP, by adjusting the displacement of one of the prisms in the hybrid compressor (see Fig. 3.1). The amount of glass inserted in the beam provides the necessary variable positive dispersion to correct the CEP by balancing the difference in group and phase velocity accordingly. Since the fundamental spectrum after the HCF is already spanning an optical octave (see Fig. 3.2), no additional spectral broadening of the branched-off pulses, e.g. in a sapphire plate or photonic crystal fiber, is necessary which may introduce artificial phase noise into the f-to-$2f$ feedback loop.

Under optimal conditions, the combined action of both control loops allows locking the CEP with an accuracy of $\sim$100 mrad (root-mean-square) [81]. However, the absolute value of the stabilized CEP remains unknown. In the actual attosecond streaking experiments, the best CEP for the production of isolated attosecond pulses is identified by observing the modulation and/or intensity of the HH cut-off radiation [68, 4]. It should be noted that very recently single-shot characterization of the CEP has been demonstrated [83]. The detection principle, which exploits the waveform-dependent asymmetry of above-threshold photoelectron emission [84], may be implemented in future attosecond experiments for online-monitoring of the CEP. This may eliminate the need for active stabilization which often provides only limited long-term stability.

3.2 Apparatus for Attosecond Electron Spectroscopy in Condensed Matter

Within the scope of this thesis a new apparatus for attosecond spectroscopic studies under ultra-high vacuum (UHV) conditions has been developed and put into operation. The whole setup was designed to provide full flexibility with respect to future application of attosecond photoelectron spectroscopy [85]. Key features include the possibility for angular-resolved studies, compatibility with all common surface science techniques for sample preparation and characterization, unlimited choice of the solid-state sample materials to be investigated (ranging from refractory metals, semiconductors to rare-gas solids), and the opportunity to perform gas-phase streaking experiments. A brief overview of the experimental setup and its most essential components will be given in this section. More technical details on the beamline and the UHV system can be found in [86].

3.2.1 The NIR–XUV Beamline

After leaving the laser system described in the previous section, the intense nearly Fourier-limited NIR pulses enter a rough vacuum system ($\sim 10^{-4}$ mbar) in which they are guided to the beamline for high-harmonic generation (HHG) only by a set of silver mirrors. In this way, the short pulse duration can be preserved and is not affected by dispersion occurring in air or other optical components. A schematic overview of the beamline is depicted in Fig. 3.5. High-harmonic radiation is generated by focusing the NIR pulses with a concave silver mirror FM (ROC[1] = -1100 mm) into a quasi-static gas target (T) backfilled with neon. The z-folded geometry with the flat silver folding mirror (M1) allows for a small angle of incidence on FM and therefore reduces aberration of the NIR focus. The gas target consists of a thin nickel tube (Ø 3×0.1 mm) in which the necessary holes are drilled by the focused laser beam itself to minimize the gas flow into the surrounding vacuum system. It is mounted on two translation stages for precise positioning along and perpendicular to the laser beam path.

In order to improve phase-matching conditions for HHG in the cut-off region, the interaction length is further reduced by squeezing the nickel tube to a thickness of $2 - 1.5$ mm in the laser propagation direction [87]. In practice, the Ne gas pressure, the target position with respect to the NIR focus and the intensity of the NIR driving pulses (controlled by aperture A1) are optimized to generate the highest photon flux in the cut-off region within the narrowest spatial mode possible (see e.g. lower inset in Fig. 3.10). Best results were obtained with Ne backing pressures between $150 - 250$ mbar. The maximum NIR intensity delivered to the gas target is limited by the ROC of the focusing mirror FM which can in principal be varied between -600 mm and -1800 mm, resulting in peak intensities of $\sim 2 \cdot 10^{14} - 2 \cdot 10^{15}$ W/cm^2. This compatibility is achieved by flexible bellows installed in the beamline allowing to adjust the distance between the FM-mount and the HHG target accordingly.

Most of the gas load emerging from the HHG target is pumped by a 600 l turbomolecular pump (Leybold MAG W 600) directly attached to the chamber accommodating the gas cell. In this way, the pressure in the HHG chamber can be kept below 10^{-2} mbar during operation which minimizes re-absorption of the generated XUV radiation [38]. A second differential pumping stage, consisting of a skimmer (S) with an opening of Ø 1.5 mm and a 200 l turbo pump, further restricts the gas flow downstream to the experimental area.

---
[1]Radius of Curvature

Apparatus for Attosecond Electron Spectroscopy in Condensed Matter

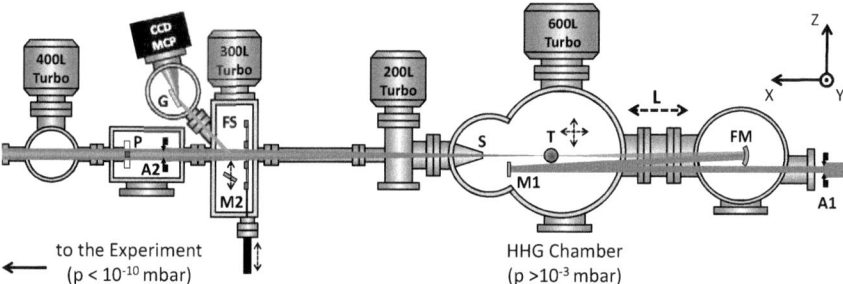

Figure 3.5: Schematic top-view of the beamline for collinear generation of few-fs NIR and sub-fs XUV pulses by high-harmonic generation. Essential components are illustrated: adjustable, motorized iris (A1 and A2), folding mirror (M1), focusing mirror (FM), gas target (T), skimmer (S), gold mirror (M2), filter slider (FS), grating (G) and the pellicle-metal filter assembly (P). Their degrees of freedom for positioning are indicated by dashed arrows. The transition from rough vacuum in the HHG chamber to UHV conditions in the experimental area is achieved by several differential pumping stages (turbomolecular pumps) incorporated in the beamline. See text for further details.

Following this pumping stage, a narrow beam pipe in combination with a 300 l turbo pump is already sufficient to maintain a pressure of $5 \cdot 10^{-9}$ mbar. The final transition to UHV with $p < 10^{-10}$ mbar is achieved by differential pumping with a 400 l turbo pump installed at the end of the beamline. This last pumping stage does not contain any temperature-sensitive components and is therefore bakeable to 200 °C.

The generated XUV radiation can be monitored in an energy range of $\sim 60 - 250$ eV with a home-built spectrometer installed after the third differential pumping stage [86]. A grazing incidence gold-palladium mirror (M2) can be inserted into the beam path to reflect the low-divergent XUV beam onto a gold-coated flat-field grating (G) which disperses the radiation horizontally on a detector. The detection device consists of a double microchannel plate (MCP) assembly in chevron geometry with a cesium-iodide coating for enhanced XUV sensitivity, and a phosphor screen on the rear side for optical detection with a CCD camera. To avoid saturation of the detector, the NIR light is suppressed by a 950 nm zirconium foil permanently installed in front of the MCP assembly.

Additional spectral filters supported on a slider feedthrough (FS) can be brought into the beam path: the $L_{2,3}$ absorption edge of free-standing foils of 150 nm thick silicon and 200 nm thick aluminum are used for photon energy calibration [88], whereas a 200 nm thick palladium (Pd) or zirconium (Zr) foil is used to completely block the NIR laser light for alignment and optimization purposes of the XUV beam in the experimental area further downstream. A typical XUV spectrum recorded with parameters optimized for efficient HHG in the energy range > 100 eV is depicted in Fig. 3.6. The CEP of the driving NIR pulses was optimized to produce the smoothest spectrum near the cut-off. Obviously, the radiation features a robust continuum supporting isolated sub-fs XUV pulses in a photon

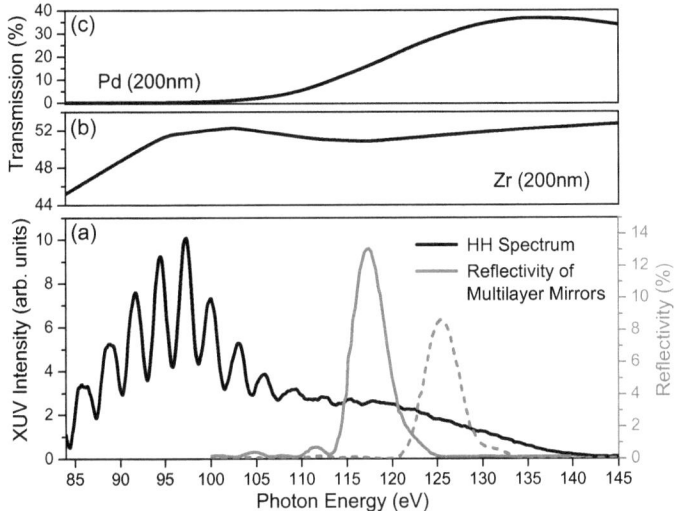

Figure 3.6: Prerequisites for the generation of isolation sub-fs XUV pulses by spectral filtering of high-harmonic (HH) cut-off radiation. (a) High-energy part of the harmonic radiation (black line) generated by focusing phase-stabilized few-cycle NIR pulses with an intensity of $\sim 6 \cdot 10^{14}$ W/cm^2 into a target supplied with $\sim$200 mbar neon gas. The carrier-envelope phase was optimized to produce the spectrum with minimal spectral modulation above 110 eV. Grey lines (dashed and solid) represent the reflectivity profiles of the multilayer mirrors employed in the majority of streaking experiments performed in this thesis. (b,c) XUV transmittance of the metal filters employed to spatially separate the cut-off radiation from the fundamental NIR light.

energy range of $\sim$100 − 135 eV.

Leaving the gas target, the generated high-harmonic beam propagates co-linearly with the remaining NIR light towards a filter assembly (P) where they are spatially separated based on their radically different divergence [35, 36]. The filter assembly consists of a circular metal foil (Ø 5 mm) suspended in a NIR-transparent nitrocellulose pellicle. The metal foil, either 200 nm thick Zr or Pd, blocks the central portion of the NIR beam and functions at the same time as a high-pass filter for the HH radiation. The transmission characteristics of both metal foils is shown in Fig. 3.6 (b) and (c), respectively. The surrounding pellicle (thickness: 5 μm) is almost transparent to the more divergent NIR driver pulses. The whole filter assembly is mounted in a sealed translation sleeve which can be moved in the plane perpendicular to the beam path for proper alignment with respect to the XUV beam. Directly in front of the pellicle-metal filter, a motorized iris (A2) on a three-axis manipulator allows to control the NIR intensity delivered to the experiment.

Apparatus for Attosecond Electron Spectroscopy in Condensed Matter

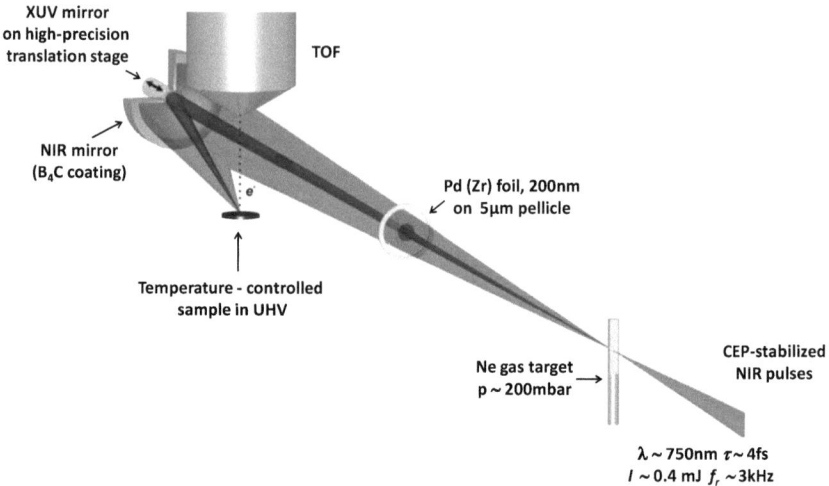

Figure 3.7: Simplified overview of the setup used for attosecond streaking experiments of surfaces [22]. High-harmonic radiation is generated by exposing neon atoms to intense, waveform-controlled few-cycle NIR laser pulses (gray beam). A thin metal filter spatially separates the low-divergent XUV radiation (dark gray beam) from the residual NIR light. The two collinear beams are reflected by a two-component mirror. The outer part of this mirror is fixed and focuses the NIR pulses onto the sample while the inner part functions as a bandpass reflector and filters isolated sub-fs XUV pulse from the HH cut-off continuum. It is attached to a piezo-electric translation stage allowing to introduce a delay between the XUV and NIR pulses. Both pulses are spatially and temporally overlapped on the sample surface positioned in the focus of the double mirror assembly. In the actual measurement, the kinetic energies of the XUV-induced photoelectrons emitted along the surface normal are analyzed by a time-of-flight spectrometer (TOF) as a function of the relative delay between the NIR and XUV pulses.

3.2.2 Selection of Isolated sub-fs XUV Pulses

After their spatial separation, the annular NIR beam and the transmitted central XUV beam enter the experimental UHV end station (see next subsection) where they are reflected by a coaxial assembly of two spherical mirrors (ROC = -250 mm) and focused onto the surface of the solid-state sample (see Fig. 3.7). The inner mirror is coated with a dedicated multilayer structure acting as a bandpass filter for the incident XUV radiation [32, 33]. The measured reflectivity of the multilayer mirror employed in most of the experiments presented in Chapter 4 is depicted in Fig. 3.8 (a). A Gaussian fit yields a central energy of $\hbar\omega_x \approx 118$ eV and a bandwidth of $\Gamma_x = 4.2$ eV. In combination with a proper HH cut-off continuum as shown e.g. in Fig. 3.6, this multilayer mirror supports isolated XUV pulses with a Fourier-limited duration of $\tau_x = 435$ asec, as determined by

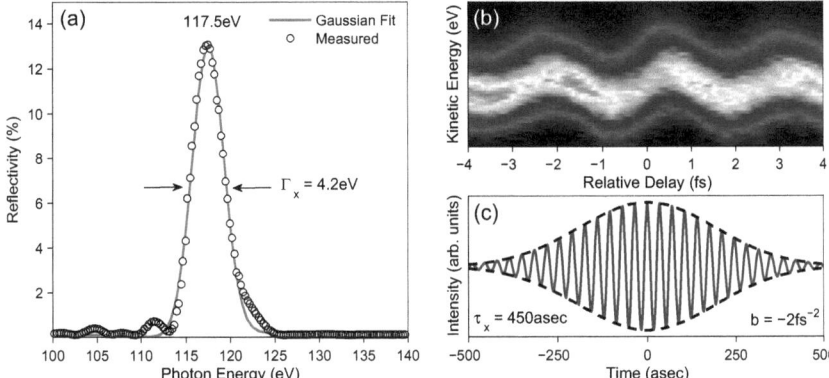

Figure 3.8: Characterization of the XUV multilayer mirror used in most of the streaking experiments reported in this thesis. (a) Reflectivity curve of the multilayer structure measured by x-ray reflectometry at the synchrotron facility BESSY-II (circles). A Gaussian function fitted to the data is shown as red line. (b) Streaking of the Xe4d photoelectrons in gas-phase xenon. The (negative) chirp, clearly discernible in the spectrogram, indicates the formation of non-transform-limited XUV pulses. (c) Most probable temporal evolution of the filtered XUV pulses.

the time-bandwidth product for Gaussian pulses [75]:

$$\tau_x[\text{asec}] \cdot \Gamma_x[\text{eV}] \geq 1825 \tag{3.11}$$

Indeed, gas-phase streaking measurements performed with HH radiation filtered by this multilayer mirror (see Fig. 3.8 (b)) confirm the generation of isolated XUV pulses with negligible satellite content. However, they also revealed the presence a negative chirp carried by the produced XUV pulses, which is most likely caused by residual distortions of the spectral phase introduced by the employed aperiodic multilayer design [33]. Comparison of the streaking spectrograms with simulations[2] (comp. Section 2.2.1) suggest a linear chirp rate of $b_x \approx -2$ fs^{-2} which, however, induces only a marginal temporal broadening of the XUV pulses to $\tau_x = 450$ asec [75]. This pulse duration is still sufficiently short compared to the period of the NIR laser pulses and does therefore not compromise the observation of the streaking effect (comp. Section 2.2). The spectral bandwidth of the filtered XUV pulses is narrow enough to unambiguously resolve the photoemission from electronic states exhibiting a separation in binding energy of ≥ 10 eV, which is crucial for the relative timing measurements performed in this thesis.

The outer mirror is coated with 13 nm of boron carbide (B$_4$C) acting as a partial reflector with a reflectivity of 20 % for the NIR pulses. The use of more conventional silver mirrors,

---

[2] An accurate analysis based on a FROG-type algorithm is not possible due to the unresolved Xe4d spin-orbit doublet.

that exhibit almost perfect reflectivity in the NIR range, turned out to be disadvantagous since it rendered the fine adjustment of the NIR intensity without exceeding the damage threshold of the solid-state samples rather difficult.

The pointing of the outer mirror is motorized, enabling spatial overlap of the focused NIR and XUV pulses. The entire double mirror assembly is mounted on a stack of 4 piezo-driven stages (Nanomotion), which allows positioning of the mirrors relative to the incident laser beam as well as steering the reflected pulses onto the sample surface. A temporal delay between the XUV and the NIR pulses can be introduced by translating the multilayer mirror along the incident beam with a closed loop piezo-electric transducer (PI Hera 621, customized for UHV), which has a full range of 100 μm and a nominal resolution of 0.2 nm, corresponding to an ultimate limit for the smallest possible delay step size of 1.3 asec. This compact concentric double mirror design constitutes a low-jitter delay stage without the need for active stabilization.

3.2.3 The Surface Science End Station

The end station consists of two cylindrical stainless steel (316 L) vessels separated by a pneumatic gate valve. One chamber is dedicated to sample preparation and characterization while the second one (experiment chamber) is exclusively dedicated for photoelectron spectroscopic studies. The longitudinal axes of both chambers are parallel. The measurement chamber is designed in an in-axis geometry, i.e. the sample, the laser focus, and the foci of all detectors in their working position coincide with the chamber axis. In contrast, an off-axis mounting was devised for the preparation chamber in a way that the foci of all the preparation/diagnostic instruments lie on a circle of 250 mm radius around the chamber axis for maximum solid-angle range of each station. The sample can be positioned in front of the individual preparation stations by rotating the manipulator along this circle by means of a large differentially pumped rotary feedthrough. Details on the design and operation principle of the rotary feedthroughs can be found in [89]. Schematic cross-sectional views of the end station are shown in Fig. 3.9 and Fig. 3.10. The preparation chamber accommodates standard equipment for surface preparation and controlled layer growth: an ion sputter gun, a four-grid back-view LEED system, a Knudsen-type effusion cell, a modified Bayard-Alpert pressure gage for temperature programmed desorption (TPD) studies, and a gas dosing system composed of a multi-capillary array doser attached to a gas manifold via a flow-controller.

The solid-state samples are mounted in an exchangeable holder attached to the head of a home-built $xyz-$manipulator by a simple clamping mechanism. A differentially pumped feedthrough at the other end of the manipulator enables 360° rotation of the sample around the $z-$axis. Fast transfer of the sample holder from atmosphere to UHV is provided by a dedicated home-built load-lock system. The sample temperature is monitored by a thermocouple (typically C-type) directly spot-welded to the edges of the sample. The reference junction of this thermocouple is attached to the sample holder,

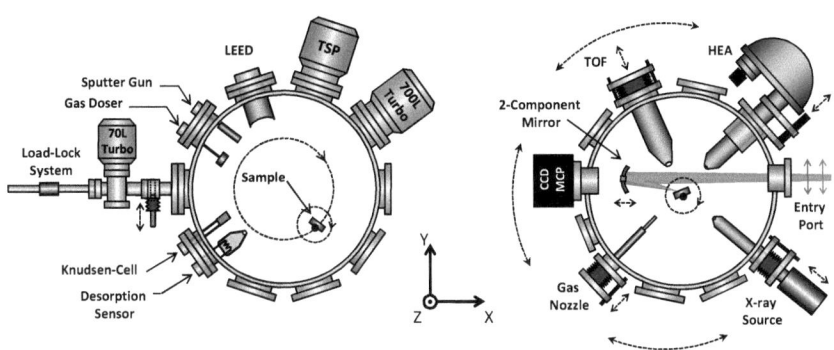

Figure 3.9: Cross-sectional view of the preparation chamber (left) and experiment chamber (right). The various degrees of freedom for positioning of the individual components are indicated by dashed arrows.

the temperature of which is measured independently with a type-K thermocouple. This design circumvents the use of specific thermocouple-vacuum-feedthroughs and therefore allows an independent choice of the sample thermocouple material, which can be adapted to the individual needs of the experiment. An available sample temperature range of 15 – 2500 K is achieved by thermal contact of the sample holder to a continuous flow-cryostat (operated with liquid helium or nitrogen) integrated in the manipulator, and by radiative and electron-beam heating of the sample, respectively. Arbitrary temperature profiles can be generated by software with a proportional-integral-derivative (PID) controlled feedback loop driving the heating power supply.

The experiment chamber is composed of a fixed section, and a detector ring that can be rotated around its longitudinal $z-$axis, enabled by two differentially pumped rotary feedthroughs. The rotating part accommodates a commercial electron time-of-flight (TOF) spectrometer (Stefan Kaesdorf, Geräte für Forschung und Industrie), a hemispherical electron energy analyzer HEA (SPECS PHOIBOS 100) equipped with a CCD detector, and a X-ray source with a Mg/Al-twin anode (PSP Vacuum Technology) for sample characterization with standard XPS[3]. This flexible configuration, combined with the sample rotation allows an independent selection of both the XUV/NIR angle of incidence and the electron detection angle. Considering the linear polarization of both the NIR and the XUV pulses, this arrangement can be used to distinguish polarization and angular effects in attosecond photoemission. For all streaking measurements presented in this thesis, the TOF spectrometer was employed to collect the photoelectrons due to its superior detection efficiency. More details on the conversion and calibration of TOF data are given in Appendix A. Gas-phase streaking measurements are enabled by a gas nozzle mounted on a $xyz-$manipulator which can be brought to the focus and replaces

---

[3]X-ray Photoelectron Spectroscopy

Apparatus for Attosecond Electron Spectroscopy in Condensed Matter

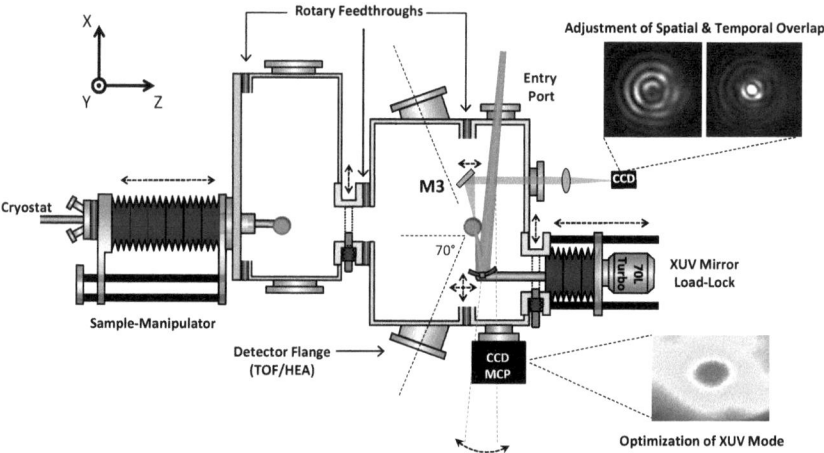

Figure 3.10: Cross-sectional side-view of the UHV end station. The upper inset shows interference patterns generated by NIR pulses reflected off the inner and outer part of the two-component mirror when their spatial and temporal overlap is perfectly adjusted. The lower inset depicts a typical spatial intensity profile observed for high-harmonic radiation near the cut-off ($\hbar\omega_x \geq 100$ eV).

the solid sample. All devices are equipped with translation and tilt mechanisms enabling their optimal alignment with respect to the sample surface.

The NIR and XUV beam enter the end station through the fixed part of the experimental chamber and are focused by the double mirror assembly onto the sample (see Fig. 3.10). For optimization and alignment purposes, the fixed part also houses an MCP/fluorescent screen system to image the spot profile of the HH cut-off radiation arriving at the experiment. In normal operation, this detector is obscured by the double mirror assembly. Therefore, the entire end station can be rotated in the xz−plane around the entry port by means of air pads, to steer the XUV beam onto the MCPs. The entire double mirror stack is mounted on a transfer system which can be retracted from the main chamber into a load-lock. In this way, the exchange of multilayer mirrors is possible without breaking the vacuum in the main UHV chamber and further enables a separate bake-out of the double mirror stage which can only withstand temperatures below ∼100 °C because of the sensitivity of the installed piezo-electric elements.

Both chambers, as well as the rotary feedthroughs, are individually pumped by a system of turbomolecular and scroll pumps. Additionally, both chambers are equipped with liquid nitrogen cooled titanium sublimation pumps (TSPs). The residual gas composition can be monitored by a quadrupole mass analyzer. After proper bake-out of the entire UHV system, and with the TSPs running, a base pressure of $7 \cdot 10^{-11}$ mbar can be routinely achieved and maintained during the measurements.

Temporally and spatially overlapping the XUV and the NIR pulses on the surface of the sample is crucial for successful streaking experiments. This is accomplished by observing the optical interference between NIR pulses reflected from both the inner and outer part of the double mirror. For this purpose, the sample is retracted from the measurement position. The NIR light reflected from both parts of the double mirror is directed by a silver mirror (M3) through a view port of the end station, and is finally imaged with a lens onto a CCD camera (see Fig. 3.10). The motorization of the outer mirror is used to adjust the spatial overlap of the two beams in the focus. In the range of temporal overlap between the two pluses, interference fringes are observed. To ensure that the wave fronts of the reflected pulses are parallel, the pointing of the outer mirror is optimized to make the interference fringes as uniform and symmetric as possible. Typical constructive and destructive interference patterns observed for a perfectly aligned double mirror are shown in the upper inset of Fig. 3.10. Due to the much shorter wavelength, the focus size of the XUV radiation is much smaller compared to the NIR focus. Therefore, this alignment procedure also guarantees that the reflected attosecond pulses are safely locked inside the NIR focus of the outer mirror. The sample is brought into this focus by adjusting the distance between the sample and the double mirror to generate the most energetic ATP electrons. The production of these electrons depends sensitively on the NIR intensity and therefore also on the NIR spot size.

According to the discussion in Section 2.2, the strongest streaking effect can be expected when the XUV-induced photoelectrons are detected parallel to the polarization direction of the NIR dressing field. However, geometric constraints imposed by the restriction to keep the angle of incident on the double mirror $< 5°$ to minimize astigmatism, and the need to keep the dimensions of the end station reasonable, made it necessary to use a tilted configuration for the detector flanges in order to allow access to the sample surface. For the TOF detector, this results in an angle of $\gamma = 70°$ with respect to the chamber $z-$axis (see Fig. 3.10). This is compensated by a corresponding tilted mount of the sample to allow the detection of electrons emitted along the surface normal. For the $\vartheta = 15° - 20°$ gracing incidence of the NIR/XUV radiation on the surface, which was employed in all streaking measurements, this gives rise to a component of the NIR electric field parallel to the surface of $E_\parallel = E \sin \alpha \approx 0.45 E$, where the angle α between the NIR polarization vector and the surface normal is given by $\cos \alpha = \cos \vartheta \sin \gamma$. In combination with the finite acceptance angle of the TOF analyzer, this may contribute some additional broadening to the streaked electron emission (comp. Fig. 2.3 in Section 2.2.1) [6, 90, 7].

3.3 Synchrotron Experiments

Unavoidably, the attosecond XUV pulses produced by HHG come with a large bandwidth, which translates into a poor spectral resolution of the resultant photoelectron spectra. On the other hand, a detailed complementary characterization of the photoemission in the spectral domain turned out to be important for the analysis and interpretation of

Synchrotron Experiments

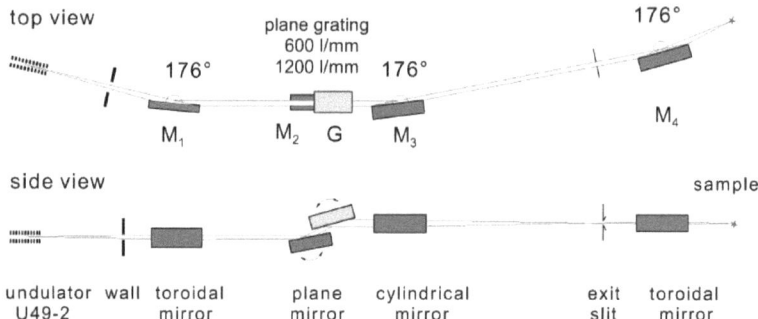

Figure 3.11: Optical layout of the undulator beamline U49/2-PGM1 at BESSY-II [91].

attosecond streaking experiments. Ideally, this characterization should be performed with photon energies similar to the central energy of the filtered attosecond pulses in order to allow reliable comparison in terms of cross-sections and character of the probed initial and final states, which both can be strongly dependent on the excitation energy. Since laboratory sources like discharge lamps and X-ray tubes do not cover the relevant XUV energy range, the electronic properties of the systems investigated in this thesis have been additionally analyzed by means of high-resolution photoemission spectroscopy at the undulator beamline U49/2-PGM1 of the third-generation synchrotron facility BESSY-II in Berlin.

An undulator is an insertion device that consists of periodically aligned dipole magnets with alternating polarity that forces the passing relativistic electrons on a sinusoidal beam path. The electric fields emitted by one electron in the different periods of the undulator interfere constructively resulting in a radiation of much higher brilliance compared to light produced by a simple bending magnet. Because the motion of the electrons in the magnetic array is not perfectly sinusoidal, the undulator radiation also contains contributions of higher (odd) harmonics which can also be used. The optical layout of the beamline is shown in Fig. 3.11. The radiation emitted from the undulator enters a plane grating monochromator (PGM) that only transmits photons within a narrow energy band depending on the position of the grating (G) and the selected size of the exit slit. The beam is finally collimated and directed onto the sample by a toroidal refocusing mirror (M4). The polarization of the photons is nearly 100 % linear in the plane defined by the storage ring. More detailed information on synchrotron radiation and insertion devices can be found in [92].

The UHV chamber used in these measurements is described in [89]. It features similar capabilities for sample preparation and diagnostics as the end station discussed in the previous section. Photoelectron spectra were acquired with p-polarized radiation under 7° gracing incidence. The photoelectrons are collected in normal emission with a hemi-

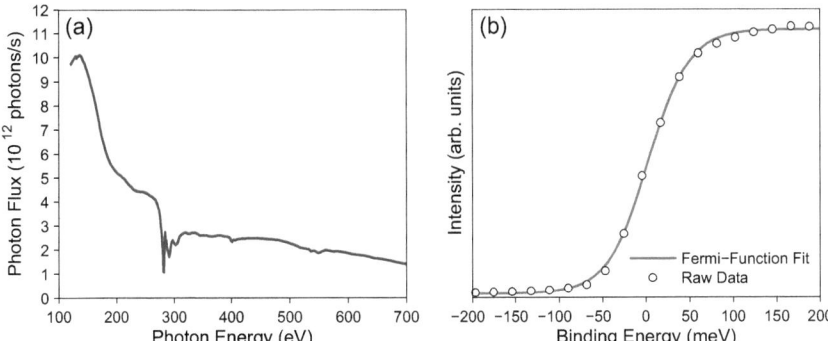

Figure 3.12: Typical conditions in synchrotron photoemission experiments. (a) Photon flux arriving at the sample measured with a type-calibrated GaAsP diode. The pronounced drop in the photon flux between $\hbar\omega = 280 - 300$ eV is caused by carbonaceous contamination on the optical elements in the beamline. (b) Photoemission from the Fermi edge of a Au(111) crystal at 90 K. The fit of the raw data to a step function convoluted with a Gaussian reveals the total instrumental resolution of 35 meV (FWHM).

spherical electron energy analyzer. A base pressure of $4 \cdot 10^{-11}$ mbar could be maintained during all measurements. The energy scale was calibrated individually for each spectrum by recording the photoemission from the Fermi edge. All intensities are normalized to the incident photon flux measured with a type-calibrated GaAsP diode (see Fig. 3.12 (a)) [93]. Only this procedure allows quantitative comparison between different spectra with highest accuracy. The available photon flux at $\hbar\omega \approx 100$ eV is almost 10^{13} photons/s. For comparison, the corresponding average XUV photon flux delivered by state-of-the-art 3 kHz high-harmonic sources, with typically $\sim 10^6$ photons per pulse in the cut-off region [94], is only $\sim 10^9$ photons/s. The total instrumental resolution for the synchrotron photoemission experiments was inferred from the broadening of the Fermi edge and can be well approximated by a Gaussian function with a FWHM of 35 meV.

In the normal multi-bunch operation mode at BESSY-II, 360 electron bunches with an energy of 1.7 GeV are stored in the ring for typically 7 h, after which losses due to electron-electron scattering and collisions with residual gas molecules require the injection of new electron packages. The produced light flashes have a duration of $\sim$20 picoseconds.

Synchrotron Experiments

Chapter 4

Surface Electron Dynamics probed by Attosecond Streaking

The availability of isolated sub-fs pulses in the XUV spectral range in combination with a precisely timed electric field of a few-cycle NIR laser pulse forms the basis for exploring electron dynamics on the attosecond times scale ($1\,\mathrm{asec} = 10^{-18}$ s). These pulses have been successfully used in various configurations to study sub-fs electron dynamics. However, all these experiments have been limited to rather simple molecules or atoms in the gas phase [11, 44, 10]. The capability of these tools for probing electron dynamics in condensed matter systems on a similar time scale was only recently demonstrated by extending the attosecond streaking technique to solids. A first proof-of-principle experiment revealed that XUV-induced photoelectrons emitted from a tungsten surface feature an intriguing temporal structure which is determined by the motion and interaction of the photo-excited electrons prior to their escape from the solid. Using the NIR electric field as a probe, Cavalieri et al. measured a relative delay of 110 ± 70 asec between the release of conduction band and W4f core-level photoelectrons [22]. However, the complexity of both the employed spectroscopic technique and the many-body interactions inherent to photo-excitation and electron propagation in condensed matter complicate a conclusive explanation for these phenomena so far [61, 63, 59, 60, 95]. Within the framework of this thesis, attosecond streaking has been applied to various prototypical surface science systems to explore the potential and limitations of this spectroscopic method. The main emphasis was to shed light on the origin of the time delays occurring in laser-dressed attosecond photoemission, which is the prime observable in these experiments [22, 44]. In this chapter, the results obtained by attosecond time-resolved photoemission from different metal surfaces as well as several well-defined surfaces-adsorbate systems exhibiting different coupling strengths are presented. It will be demonstrated that under favorable conditions temporal effects in solid-state photoemission can be followed with a precision approaching 10 asec. This enhanced accuracy allows the observation of subtle adsorbate-induced effects and permits tracking the motion of photo-excited electrons in condensed matter systems in real time as they traverse single atomic layers.

4.1 Attosecond Photoemission from Clean Metal Surfaces

Clean metal surfaces are probably amongst the simplest model systems for solid-state photoelectron spectroscopy. Given the rather low photon energy of ≤ 140 eV at which sub-fs pulses can be produced with sufficient efficiency by state-of-the-art high-harmonic (HH) sources, the application of attosecond streaking is currently limited to materials featuring shallow core-levels with large photo-ionization cross-sections. In this section, results from attosecond streaking experiments performed on the (110) surface of tungsten and the (0001) surface of magnesium are compared. Both metals exhibit suitable core levels which can be conveniently excited with photon energies in the available HH cut-off continuum. The W(110) surface will serve as a benchmark system to verify the existence of time delays in attosecond photoemission from conduction band and core states in solids, and to assess the achievable accuracy of these relative timing measurements in surface streaking experiments. The alkaline-earth metal magnesium is a prototype of a simple metal with nearly free-electron-like character of its electronic states. Especially, its conduction band should comply better with the jellium approximation, which has been previously applied to model attosecond photoemission from surfaces [61, 59]. In this respect, it differs significantly from the transition metal tungsten where the rather atomic-like $5d$ electrons dominate the electronic properties. Contrasting results obtained from these two different metals may therefore help to elucidate the mechanisms behind the temporal characteristics of attosecond photoemission. Special attention was also paid to a possible influence of the surface condition on the experimental results.

4.1.1 The (110) Surface of Tungsten

After transferring the W(110) single crystal from ambient to UHV, an atomically clean surface is produced by Ne$^+$ sputtering and repeated cycles of oxidation and annealing from $400-2200$ K in $1 \cdot 10^{-7}$ mbar pure oxygen atmosphere. This treatment efficiently removes carbon impurities from the tungsten surface [96]. The resultant oxygen layer effectively passivates the surface against further adsorption of contaminants from the residual gas. The final cleaning step consists of several rapid high temperature flashes up to 2400 K in UHV to desorb excess oxygen as tungsten oxide and to obtain a smooth surface with a low defect density [97]. Crystals prepared in this way exhibited sharp LEED patterns (see e.g. Fig. 4.20 (b)) and showed no evidence of adsorbed or segregated impurities in XPS. This established preparation procedure was applied on a day-to-day basis while during the experiments short temperature flashes to 2400 K were sufficient to restore a clean and well-ordered surface. All experiments presented in the following sections were conducted with the sample at room temperature.

A photoelectron spectrum obtained from a freshly cleaned W(110) surface with $\hbar\omega =$ 120 eV synchrotron radiation is shown in Fig. 4.1 (a). The spectrum is dominated by

Attosecond Photoemission from Clean Metal Surfaces

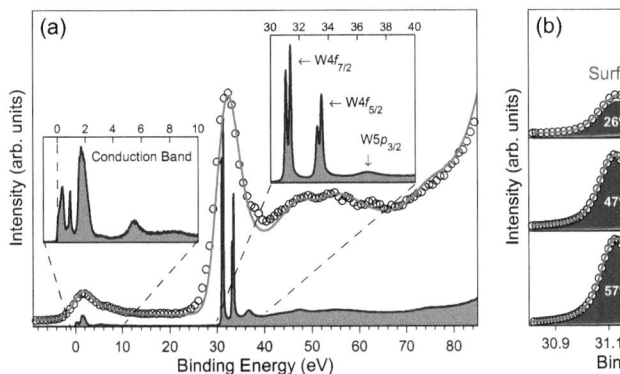
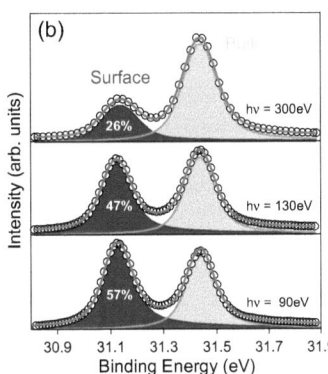

Figure 4.1: Stationary photoemission from atomically clean W(110). (a) Comparison of high-resolution synchrotron excitation at $\hbar\omega = 120$ eV (gray shaded area) to the same spectrum obtained with HH radiation filtered by the 4.2 eV FWHM multilayer bandpass reflector centered at ~118 eV (circles). The solid line corresponds to the synchrotron spectrum convoluted with a Gaussian function (4.4 eV FWHM). Insets emphasize the spectral fine structure in the conduction band and W4f emission. (b) Narrow bandwidth synchrotron radiation is used to track the surface-to-bulk ratio in the W4$f_{7/2}$ emission as a function of photon energy. The evolution of this ratio is a consequence of the kinetic energy-dependent escape depth of photoelectrons in tungsten.

the emission from the spin-orbit split W4f states appearing in a binding range of 30 − 34 eV. Each constituent of the doublet is further split into a surface and a bulk component which are separated by only 320 meV (comp. Fig. 4.1 (b)). The bulk component comprises the photoemission from all sub-surface layers with their intensities weighted by the escape depth. A quantitative decomposition of the W4f emission in surface and bulk contributions is accomplished by fitting the spectrum with two spin-orbit split doublets of Doniach-Šunjić functions [98] (convoluted with a Gaussian to account for the instrumental resolution and phonon broadening) over an energy range wider than shown in Fig. 4.1 (b) and including the W4$f_{5/2}$ emission. Fit parameters like lifetime broadening, asymmetry parameter, and spin-orbit splitting are chosen in accordance with [99] leaving only the surface-to-bulk intensity ratio free for optimization. This analysis reveals that for excitation with 120 eV photons almost 50 % of the emitted W4f electrons originate from the first atomic layer thereby emphasizing the extreme surface sensitivity in this photon energy range. Exemplary fits for different excitation energies are also shown in Fig. 4.1 (b). The decreasing surface contribution for higher photon energies is inextricably related to the kinetic energy dependency of the electron inelastic mean free path in tungsten.

The electron energy loss function of tungsten exhibits its largest amplitudes in the energy range of 15 − 30 eV, mainly due to the excitation of surface and bulk plasmons [100]. The corresponding loss lines accompanying the primary W4f emission form a broad spectral feature at ∼50 eV binding energy which is partly overlapping with the weak W$5p_{1/2}$ photoemission line at ∼47 eV. Next to the W$4f_{5/2}$ photoemission peak, at a binding energy of 36.6 eV, the W$5p_{3/2}$ emission line contributes to the spectrum with an intensity of only 7 % relative to the total W$4f$ emission strength. The conduction band of W(110) features sharp structures arising from the narrow $5d$-band with the contribution of the $6sp$-derived states being negligibly small [101]. It is also known that conduction band photoemission from the W(110) involves a significant amount of transitions related to surface states and resonances [102, 97, 103, 104]. This surface emission may be enhanced by the p-polarization of the exciting radiation [56]. In this context it should be remembered that both the NIR and the XUV radiation employed in the streaking experiments are also predominately p-polarized.

A photoelectron spectrum of clean W(110) recorded with sub-fs XUV pulses filtered from the HH cut-off radiation with the 4.2 eV bandpass reflector centered at ∼118 eV is shown as circles in Fig. 4.1 (a). For this broad bandwidth excitation, all the fine structure in the photoemission is lost and has merged into two main peaks with slightly asymmetric tails to higher binding energies. Nevertheless, these two broad spectral features remain clearly separated and distinguishable from the background which is a prerequisite for probing their temporal characteristics by attosecond streaking. In the following, the broad feature centered at ∼32 eV binding energy will simply be referred to as the W$4f$ peak, which is justified considering the small contribution of the overlapping W$4p_{3/2}$ peak at this photon energy. The full spectrum is well reproduced by convolving the synchrotron spectrum with a 4.4 eV (FWHM) broad Gaussian (red line in Fig. 4.1 (a)). This FWHM is slightly larger than the bandwidth of the XUV mirror used to filter the HH cut-off radiation (comp. Fig. 3.8 (a)), mainly because of the finite energy resolution of the TOF analyzer (see Appendix A).

The mere fact that the line-shapes in the photoelectron spectrum obtained with sub-fs XUV pulses can be satisfactorily reproduced by a symmetrically broadened synchrotron spectrum indicates that possible interference effects in attosecond photoemission are absent, or at least, do not give rise to resolvable intensity modulations for photon energies near ∼120 eV. In principle, such interference phenomena could be expected since the energetic width of the coherent XUV radiation is larger than the 2.2 eV spin-orbit splitting of W$4f$ and similar to the energy range comprising the most intense conduction band transitions. Furthermore, the good agreement in the high binding energy region proves that the energy distribution of inelastically scattered electrons is not significantly different for sub-fs XUV excitation. This is at variance with a recent classical transport calculation for tungsten that predicts a characteristic $1/E_{kin}$-dependence of the background signal for attosecond XUV excitation [105].

Characteristic changes in the photoelectron spectra can be observed when the crystal surface is simultaneously illuminated with the intense NIR pulses. Figure 4.2 (a) compares the XUV-induced photoemission from W(110) to the same spectrum acquired in the presence of NIR laser pulses with a temporal delay of -30 fs between the XUV and NIR

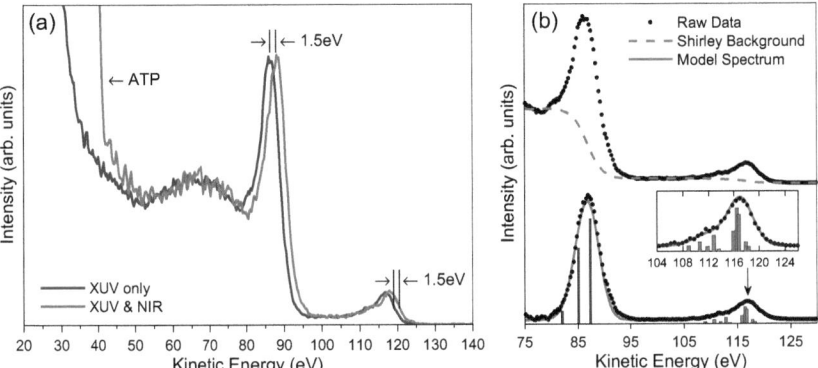

Figure 4.2: (a) Comparison of NIR field-free and NIR-dressed photoemission spectra acquired -30 fs from the temporal overlap between the XUV and NIR pulses. The NIR induced above-threshold photoemission (ATP) electrons provoke a space charge shift of the entire electron spectrum while the shape of the spectral features are preserved. (b) Top: illustration of the applied Shirley-background subtraction scheme. Bottom: the background-corrected spectrum is compared to a model spectrum constructed by summing the energy spectra of individual Gaussian electron wave packets with durations of $\tau_x = 450$ asec (comp. Eq. 4.4). Vertical bars indicate the initial-state energies associated with these wave packets and the relative weight of their corresponding energy spectrum.

pulses. The most striking effect is the appearance of a pronounced low-energy electron signal originating from conduction band electrons that have absorbed multiple laser photons from the NIR field. Since the work function ϕ of the W(110) surface is 5.25 eV [12], at least a 4-photon process is necessary to generate such above-threshold photoemission (ATP) electrons in tungsten. For typical NIR intensities of $\sim 1 \cdot 10^{11}$ W/cm^2 required for surface streaking experiments, the energy distribution of these ATP electrons can extend to maximum kinetic energies of ~ 40 eV, corresponding to the absorption of ~ 24 laser photons.

The presence of the NIR field also causes an overall shift of the photoelectron spectra to ~ 1.5 eV higher kinetic energies due to space-charge effects. This space-charge shift could be clearly observed and distinguished from laser-dressing phenomena by recording spectra far from temporal overlap between the XUV pump and NIR probe pulse and comparing them to a NIR field-free measurement (see Fig. 4.2 (a)). In general, space-charge can also lead to distortions of the photoelectron spectra. This occurs when the XUV-induced photoelectrons traveling to the detector are perturbed by the inhomogeneous cloud of the slow ATP electrons excited by the intense NIR pulses [106]. However, no space-charge induced broadening or distortion of spectral features could be detected for the NIR intensity levels employed in the surface streaking experiments reported in this thesis. Recently, such distortions have been observed for NIR intensities exceeding $5 \cdot 10^{11}$ W/cm^2

in laser-dressed photoemission experiments on Pt(111) with 30 fs NIR pulses [107, 58]. Another source of spectral deformations may arise from the excitation of conduction band electrons into empty energy bands just above the Fermi level by the laser pulses [107]. When probed with the XUV radiation, this "hot" electron distribution will manifest itself in a broadening of the electron spectrum near the Fermi edge. This effect, however, is probably masked by the 4.2 eV bandwidth of the exciting XUV pulses and could therefore not be unambiguously identified in the NIR-dressed electron spectra.

Owing to the comparably high photon energy of the filtered XUV radiation, it was possible to maintain a sufficient energy separation between the ATP background signal and the W4f emission in all the streaking experiments that will be presented in the remainder of this chapter. In particular, the comparison in Fig. 4.2 (a) proves that the background for kinetic energies above 50 eV stems predominantly from secondary electrons produced by inelastic scattering of the primary XUV-induced photoelectrons inside the metal. The background in this energy range can therefore be removed by applying standard procedures developed for stationary photoemission spectroscopy [12, 108]. For the broad spectral features characteristic for sub-fs XUV excitation, the background correction scheme proposed by Shirley [109] was found to give reasonable results. If not stated differently, all photoelectron data presented in the following had been subjected to this background correction (including all laser-dressed spectra). An example of the applied background subtraction procedure is given in Fig. 4.2 (b). It has to be emphasized that background correction in photoemission spectra constitutes merely a recalibration of the intensity scale which facilitates the determination of peak shapes and relative emission strengths. However, it does of course not eliminate a possible influence of secondary electrons in the time domain. Especially in the vicinity of a photoemission line, the contribution of electrons that have undergone only a few inelastic scattering events cannot be strictly separated. The release of these electrons from the surface can nevertheless be correlated in time with the emission of the corresponding primary photoelectrons. Both types of electrons will contribute to the final electron wave packet whose temporal characteristics are probed in a streaking measurement [63].

In order to investigate the temporal evolution of the electron wave packets released from the CB and W4f states upon absorption of the sub-fs XUV radiation, a sequence of photoelectron spectra is collected while varying the relative delay between the XUV-pump and the NIR-probe pulses. A false-color representation of the resulting background-corrected spectrogram is shown in Fig. 4.3 (a). In the range of temporal overlap between the two pulses, the kinetic energies of both the CB and W4f electrons are deeply modulated by the NIR dressing field with the center of their energy distributions tracing the vector potential of the NIR laser pulse [47]. Without any further analysis, this observation already demonstrates two important aspects: the successful selection of isolated sub-fs XUV pulses by the employed multilayer mirror, and a temporal confinement of the photoelectron wave packets released from the W(110) crystal to less than a half-cycle of the dressing laser field (comp. Section 2.2). Apart from shifting the center of the photoemission peaks, the streaking field also changes their line-shapes. A closer inspection of the

Attosecond Photoemission from Clean Metal Surfaces

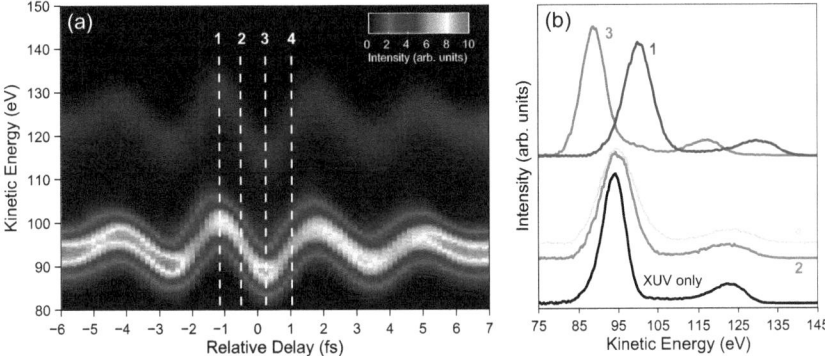

Figure 4.3: Attosecond time-resolved photoelectron emission from the clean W(110) surface. (a) Streaking spectrogram composed of 86 individual photoelectron spectra recorded as a function of relative delay τ between the NIR and XUV pulses. Each spectrum was accumulated over 10^5 laser shots in the presence of a $4.3 \cdot 10^{11}$ W/cm^2 intense NIR dressing field. A multilayer mirror acting as a 4.2 eV broad bandpass filter centered near $\sim$126 eV was used to select isolated sub-fs XUV pulses from the HH cut-off radiation. (b) Streaked electron spectra recorded at maxima (1), minima (3) and adjacent zero-crossings (2,4) of the NIR vector-potential are compared to a NIR field-free spectrum (black line). A NIR-induced broadening due to the finite temporal duration of the electron wave packets is clearly discernible.

streaked photoelectron spectra depicted in Fig. 4.3 (b) reveals a pronounced broadening of the photoelectron spectra, compared to the NIR field-free case, when recorded at relative NIR-XUV delays corresponding to the zero-transitions of the laser vector potential. This is a clear signature of the finite temporal duration of the released electron wave packets [34]. Furthermore, this broadening is slightly different when observed at adjacent zero-crossings of the vector potential, which is indicative of a linear frequency chirp of the electron wave packets (comp. Fig. 2.4 on page 14). A slight asymmetry in broadening can also be identified in spectra taken near the local maxima and minima of the vector potential, which may be ascribed to the presence of higher-order chirp [110]. However, these wave packet properties cannot simply be related to interactions between the electrons and the solid because the dominant contribution to the electron wave packet's duration and chirp is defined by the exciting XUV pulses. A detailed characterization of the filtered XUV radiation, e.g. in gas-phase streaking experiments, would be necessary to reliably disentangle these contributions. However, such a calibration has not been performed for the multilayer mirror employed in this specific measurement.

As pointed out in Section 2.2, a time delay between the release of electron wave packets from the surface should manifest itself in a relative shift of the respective parts in the spectrogram [22, 44]. Obviously, such an offset cannot be inferred from simple inspection of the raw data, even when the two spectral regions are normalized to the same intensity as in Fig. 4.5 (a). A more quantitative analysis can be based on comparing the first moments

Attosecond Photoemission from Clean Metal Surfaces

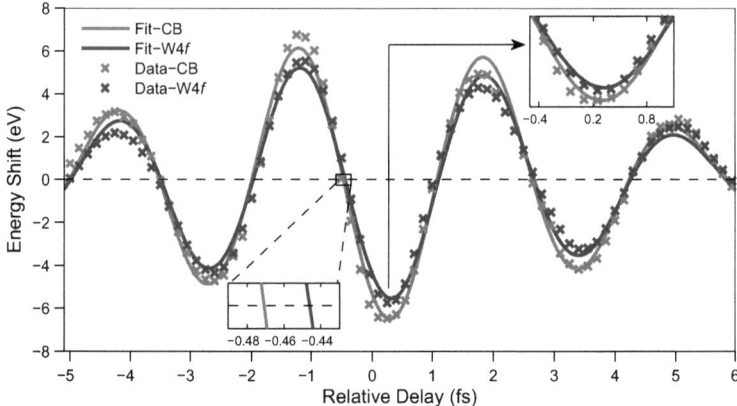

Figure 4.4: Center-of-energy (COE) analysis of the streaking spectrogram shown in Fig. 4.3 (a). Crosses represent the first moments of the photoelectron distribution calculated according to Eq. 4.1 for the conduction band and W4f region. Solid lines result from a global fit to both traces with functions describing the waveform of the NIR vector potential (comp. Eq. 4.2). The insets highlight the existence of a small relative shift $\Delta\tau \approx 25$ asec between the two parts of the spectrogram.

$S_{coe}^q(\tau)$ (also called center-of-energies (COE)) calculated for the corresponding regions of the spectrogram $P(E,\tau)$ as a function of the relative NIR-XUV delay τ according to:

$$S_{coe}^q(\tau) = \frac{\int_{E_{min}}^{E_{max}} E\, P(E,\tau)\, dE}{\int_{E_{min}}^{E_{max}} P(E,\tau)\, dE} - \tilde{E}_q. \tag{4.1}$$

Here $\tilde{E}_q$ denotes the central kinetic energy of the photoelectron peak q determined far away from the temporal overlap between the NIR and XUV pulses. In the absence of space charge effects, $\tilde{E}_q$ would be identical to the kinetic energy $E_q = E_{kin,q} = \hbar\omega_x - E_{bin,q}$ calculated from the binding energy $E_{bin,q}$ of the respective initial state q. In the following, a data set in the form of $S_{coe}^q(\tau)$ will be referred to as streaking trace of electrons released from the electronic state q.

Streaking traces for the W4f and CB photoelectrons extracted from the spectrogram in Fig. 4.3 (a) are shown as crosses in Fig. 4.4. The bounds of integration for the calculation of the COEs were set to $E_{max} = 140$ eV and $E_{min} = 110$ eV for the CB region and to $E_{max} = 110$ eV and $E_{min} = 80$ eV for the W4f levels. A possible shift $\Delta\tau$ between the streaking traces is quantified by simultaneously fitting them to analytical functions mimicking the vector potential waveform of the NIR streaking pulse:

$$\begin{aligned} S_{fit}^{4f}(\tau) &= S_0^{4f} e^{-4\ln 2\,((\tau+\Delta\tau)/\tau_L)^2} \sin\left(\omega_L(\tau+\Delta\tau)(\tau+\Delta\tau) + \varphi_{CE}\right) \\ S_{fit}^{CB}(\tau) &= S_0^{CB} e^{-4\ln 2\,(\tau/\tau_L)^2} \sin\left(\omega_L(\tau)\tau + \varphi_{CE}\right), \end{aligned} \tag{4.2}$$

where τ_L is the FWHM of the Gaussian NIR pulse envelope, φ_{CE} denotes the carrier-envelope phase[1] and a linear chirp β of the NIR pulses is taken into account by setting $\omega_L(\tau) = \omega_L + \beta\tau$. The result of this fit is shown as solid lines in Fig. 4.4. It suggests a relative time shift of $\Delta\tau \approx 25$ asec between the CB and W$4f$ streaking trace. Since for positive XUV-NIR delays the XUV pulses arrive earlier at the surface than the NIR pulses, this time shift implies a delayed emission of the W$4f$ electrons which, compared to the CB electrons, sense the same phase of the streaking field only for $\tau + \Delta\tau$ asec larger NIR-XUV pump-probe delays.

The fitted ratio of the streaking amplitudes $S_0^{CB}/S_0^{4f} = 1.18$ is in good agreement with the theoretical value of 1.15 expected from the square-root dependence of streaking amplitude on the initial electron kinetic energy (see Eq. 2.9) which further confirms the consistency of the applied background correction and data fitting scheme. The intensity I_L of the streaking field calculated from the modulation depth S_0^q according to

$$I_L = \frac{1}{2}\epsilon_0 c E_L^2 = \frac{1}{4e^2}\epsilon_0 c m_e \left(\frac{\omega_L S_0^q}{\sqrt{E_{kin,q}}}\right)^2, \quad (4.3)$$

amounts to $4.3 \cdot 10^{11}$ W/cm^2 which is still one order of magnitude below the typical damage threshold of metal surfaces [111].

Analyzing streaking spectrograms with the COE method has the advantage that no peak shape for the photoemission lines has to be specified. Although this constitutes maybe the simplest approach to identify relative time shifts in streaking measurements, it has to be verified to what extent a more realistic description of the rather complex structure of the electron wave packets emitted from the W(110) crystal influences a quantitative evaluation. The quantum mechanical description of a free electron interacting with a laser field characterized by a vector potential $A_L(t)$ leads to Volkov wave functions which arise as solutions of the time-dependent Schrödinger equation (TDSE) in the strong-field approximation [51]. According to the discussion in Section 2.2.1, an approximate representation of the time-domain wave packet for electrons excited from a discrete bound state q in the presence of a laser field is (atomic units):

$$\chi_q(t) = f_q(t)\, e^{-iE_{kin,q}} e^{i\Phi_v^q(t)} = f_q(t)\, e^{-iE_{kin,q}} e^{-i\int_t^\infty \sqrt{2E_{kin,q}}\, A_L(t') + \frac{1}{2}A_L(t')^2\, dt'}. \quad (4.4)$$

The corresponding photoelectron energy spectrum $P_q(E,\tau)$ as a function of the relative NIR-XUV delay τ is then given by the absolute square of the Fourier transform in the energy domain [22, 44, 52]:

$$P_q(E,\tau) = \left|\int_{-\infty}^{\infty} dt\, f_q(t+\tau)\, e^{i\Phi_v^q(t)} e^{i(E-E_{kin,q})t}\right|^2. \quad (4.5)$$

---
[1]To be precise, the effective streaking field at the surface is a superposition of the incoming and the phase-shifted reflected NIR field. Therefore φ_{CE} in Eq. 4.2 might differ from the carrier-envelope phase of the incident NIR pulse.

Neglecting effects of interference, the full measured spectrogram $P(E, \tau)$ can be reconstructed by superimposing the energy spectra associated with the electron wave packets launched from the different initial states of the solid:

$$P(E, \tau) = \sum_q P_q(E, \tau). \tag{4.6}$$

However, additional constraints have to be imposed to simplify such a reconstruction of experimental streaking data:

1. The envelope of all individual wave packets is assumed to be of Gaussian type:

$$f_q(t) = f_q^0 \, e^{-4\ln 2 \, (t/\tau_q)^2 + i \, b_q t^2}.$$

2. The duration τ_q and linear chirp b_q are assumed to be identical for all wave packets i.e. $b_q = b_x$ and $\tau_q = \tau_x$ for all q. This can be justified since these properties will be mainly dominated by the attosecond pulse.

3. In conformity with Eq. 4.2, the vector potential $A_L(t)$ of the streaking field is parameterized by a Gaussian envelope:

$$A_L(t) = A_0 \, e^{-4\ln 2 \, (t/\tau_L)^2} \sin\left(\omega_L t + \beta t^2 + \varphi_{CE}\right)$$

4. The values for the relative amplitudes f_q^0 of the individual electron wave packets and for the energies of the associated initial electronic states $E_{kin,q}$ are based on peak positions and transition intensities inferred from corresponding high-resolution synchrotron photoemission data (when available).

For the specific case of clean W(110), the final wave packet representing the broad W4f feature is actually composed of three individual wave packets related to the photoelectron emission from the W5$p_{3/2}$, W4$f_{5/2}$ and W4$f_{7/2}$ core states, whereas the final CB wave packet can be modeled by superimposing a minimum of 10 individual transitions. Due to their small energy separation, a further partitioning of the W4f wave packet into surface and bulk contribution is not included here to reduce the number of parameters and to prevent overfitting. As a further simplification, the individual contributions to a final electron wave packet are defined to be strictly synchronized in time, i.e. there is no relative time delay between electrons released from these states[2]. The unstreaked energy spectrum constructed from these two parametrized wave packets, assuming an XUV pulse duration of $\tau_x = 450$ asec and a central XUV energy of $\hbar\omega_x = 117.5$ eV, is shown as red line in Fig. 4.2 (b). The vertical bars indicate the relative intensities $\left|f_q^0\right|^2$ and central energies $E_{kin,q} = \hbar\omega_x - E_{bin,q}$ of the individual transitions contributing to the CB and the W4f wave packet (red: CB; blue: W4f). The reconstructed photoelectron spectrum is in excellent agreement with a measured XUV-only spectrum thereby confirming the validity of this initial-state parameterization.

---

[2]For brevity, such a group of individual, but temporally synchronized wave packets will simply be referred to as (final) electron wave packet although its individual constituents are treated incoherently when calculating the energy spectrum according to Eq. 4.6.

Attosecond Photoemission from Clean Metal Surfaces

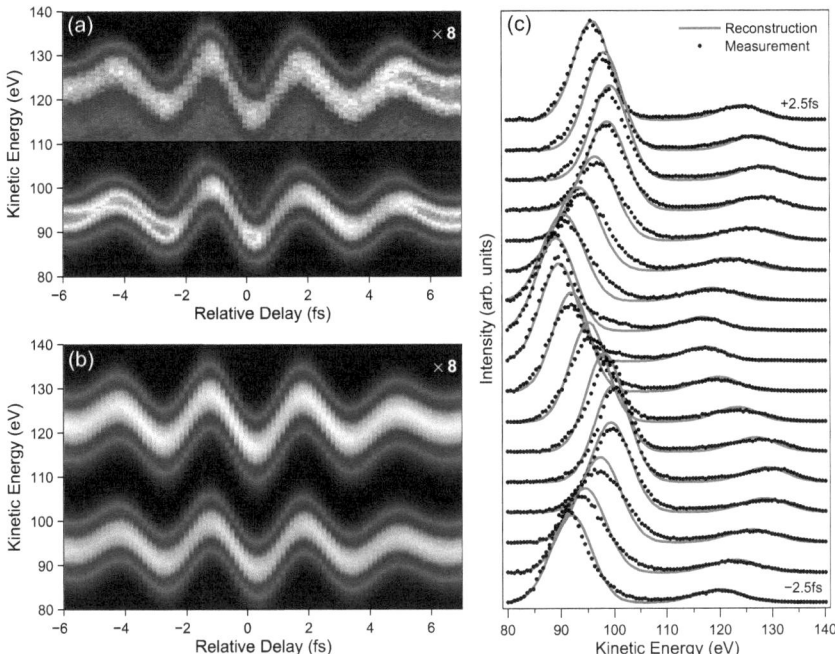

Figure 4.5: Analysis of the W(110) streaking spectrogram shown in Fig. 4.3 with a fitting procedure based on a parameterized NIR field and Volkov-type Gaussian electron wave packets (TDSE-retrieval). See text for further details. (a) Measured spectrogram. (b) Reconstructed spectrogram optimized by the TDSE-retrieval. (c) Comparison of reconstructed and measured photoelectron spectra for relative XUV-NIR delays featuring the largest kinetic energy modulation. The CB region in (a) and (b) is scaled by ×8 to simplify comparison.

In a non-linear least squares fitting routine, in the following simply referred to as TDSE-retrieval, the parameters A_0, ω_L, β, τ_L, b_x and τ_x are optimized to achieve the best agreement with the measured spectrogram while keeping the relative intensities $\left|f_q^0\right|^2$ and transitions energies $E_{kin,q}$ fixed. In analogy to the COE approach, a possible temporal delay $\Delta\tau$ between the CB and W4f wave packets is included as an additional fit parameter in the optimization scheme by substituting $\tau \to \tau + \Delta\tau$ in the calculation of the W4f energy spectrum. Figure 4.5 illustrates the result of this spectrogram reconstruction which is in reasonable agreement with the experimental data. Residual differences in line-shape between measured and reconstructed photoelectron spectra (Fig. 4.5 (c)) may be attributed to higher-order chirp of the filtered XUV pulses [110] which is not included in the fitting scheme. Such effects cannot unequivocally be distinguished from subtle line-shape variations recently predicted for attosecond streaking from solids [57]. On the other hand, the relative time delay of $\Delta \tau = 25 \pm 3$ asec between the two wave packets retrieved by this spectrogram fitting is consistent with the time shift extracted with the simpler COE approach.

Time delays of $20 - 30$ asec correspond to less than 2% of the NIR period and therefore cause only subtle shifts between the two streaking traces. Acknowledging that quantitative streaking spectroscopy is a fairly new field of research, it is indispensable to ascertain the reproducibility and overall precision of these kinds of measurements. To this end, several streaking spectrograms have been obtained from the clean W(110) surface and were analyzed with the TDSE-retrieval as described above. The evaluation of more than 50 streaking spectrograms is summarized in Fig. 4.6. A statistical analysis yields a mean value of $\overline{\Delta \tau} = 28$ asec for the time delay with a standard deviation of $\sigma = \pm 14$ asec. Even though this represents the highest absolute accuracy ever demonstrated in time-resolved solid-state spectroscopy to date, significant outlyers of so far unknown origin are nevertheless possible (see Appendix B for a discussion on possible sources of *systematic* errors in streaking experiments). Consequently, a large number of measurements are in general necessary to discriminate reliably between different timing events by means of attosecond streaking.

Within the scatter of the data presented in Fig. 4.6, no correlation can be established between the CB-W4f delay and the strength of the NIR streaking field applied in the measurements. This demonstrates the robustness of the observed effect over the whole NIR intensity range tolerable in surface streaking experiments. In principle, the NIR field penetrating the crystal can actively modify the motion of the photo-excited electrons inside the metal. Indeed, a classical transport simulation predicts a decrease of the relative delay in tungsten due to enhanced scattering of photoelectrons mediated by the residual NIR field inside the solid. On the other hand, this reduction in time shift was calculated to be only 10 asec when comparing the extreme cases of NIR field-free emission and electron emission dressed with a $2 \cdot 10^{12} \,\mathrm{W/cm^2}$ strong streaking field [63]. Obviously, the observation of such a subtle dependency would require a much higher accuracy than demonstrated in Fig. 4.6. Finally, it is also evident that the retrieved delay does not sensitively depend on the central energy ($\sim$118 eV and $\sim$126 eV) of the multilayer mirrors used to isolate the sub-fs XUV pulses from the HH cut-off continuum thereby ruling out a strong influence of final-state effects in this energy range.

The statistical analysis of the other key parameters optimized during the TDSE-fitting of the data in Fig. 4.6 yields the mean values of $\overline{\lambda}_L = 800 \pm 50$ nm, $\overline{\tau}_x = 280 \pm 33$ asec and $\overline{b}_x = -3 \pm 5$ fs^{-2} for the NIR carrier wavelength, the wave packet duration and the linear chirp rate, respectively. It should be noted that the reproducibility in retrieving these parameters was often better than 5% for spectrograms recorded shortly after each other. This points towards alignment and laser performance related influences in the streaking experiments. In spite of the large error margin, the mean linear chirp rate $\overline{b}_x$ is rather close to the corresponding value inferred from gas-phase streaking experiments (comp. Section 3.2.2), which corroborates the assumption that the dominant contribution

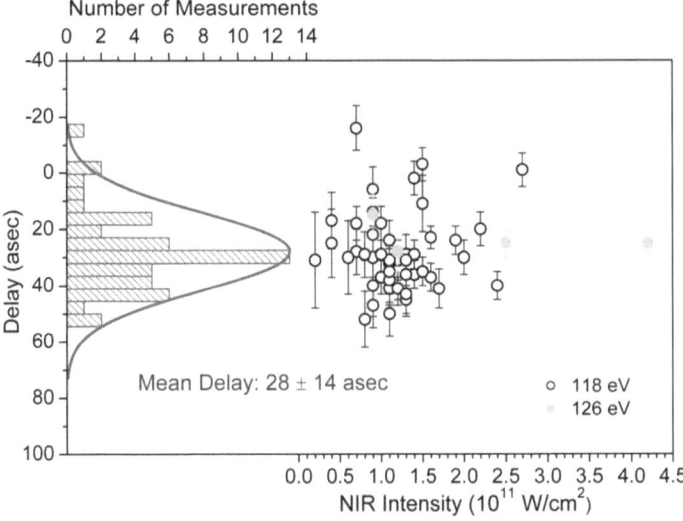

Figure 4.6: Summary of streaking measurements obtained from clean W(110). All delays are extracted with the TDSE-fitting routine. The statistical analysis reveals that photoemission from the localized W4f core states is delayed by $\overline{\Delta \tau} = 28 \pm 14$ asec with respect to the electrons photo-excited from the conduction band of the crystal. The error bars indicated in the figure correspond to the standard errors of the individual fits. In all experiments, the isolated attosecond pulses were filtered from the high-harmonic cut-off continuum by means of a multilayer mirror reflecting the incident radiation over a bandwidth of 4.2 eV (FWHM) centered either at ∼118 eV (circles) or ∼126 eV (green dots).

to the electron wave packet chirps related to the exciting XUV pulses filtered by the multilayer mirror (see Fig. 3.8 (b) on page 39). The retrieved values for τ_x are, however, not physically meaningful since they imply wave packet durations that are significantly shorter than Fourier-transform-limited XUV pulses supported by the employed multilayer mirrors (∼435 asec). Probably, the combination of a residual NIR/XUV pump-probe timing jitter, geometrical effects in the electron detection (see Section 2.2.1) and the finite energy resolution of the TOF spectrometer contribute to an additional broadening of the streaked photoelectron spectra[3]. In the time domain, this spectral broadening is misinterpreted by the fitting routine as an reduced wave packet duration. Fortunately, the retrieved time delays between the electron wave packets did not correlate with any of these parameters indicating that this relative quantity is less susceptible to existing experimental deficiencies.

Both the analysis of spectrograms of different quality and extensive tests with simulated

---

[3] A NIR-XUV timing jitter and geometric effects would not affect the line width of XUV-only spectra.

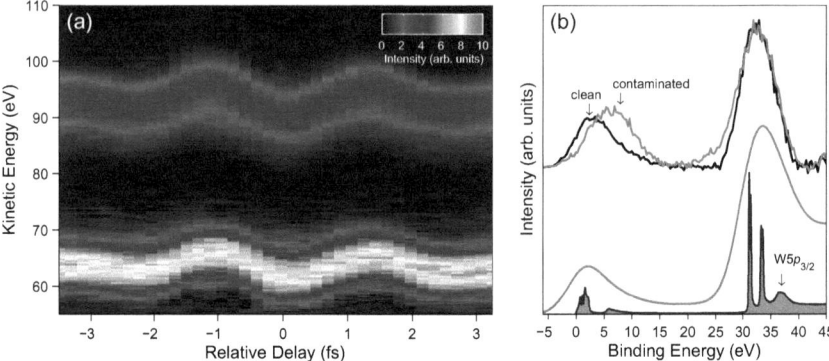

Figure 4.7: Attosecond photoemission from W(110) acquired under similar experimental conditions as reported in [22]. (a) Streaking spectrogram obtained with sub-fs XUV pulses filtered by a Mo/Si multilayer mirror reflecting incident XUV radiation at ∼94 eV over a bandwidth of 6.8 eV (supporting 270 asec Fourier-limited Gaussian pulses). (b) Top: Comparison of background-corrected NIR field-free spectra acquired from freshly cleaned W(110) (dark gray line) and the same spectrum obtained under non-UHV conditions (gray line). Bottom: Photoelectron spectrum obtained with $\hbar\omega = 94$ eV synchrotron radiation (gray shaded area). The solid line results from a convolution of the synchrotron spectrum with a 7 eV FWHM Gaussian.

streaking data confirmed that the additional constraints implemented in the TDSE-fitting routine make this method more robust and enabled a more reliable evaluation of streaking data even when recorded under non-optimal experimental conditions. Especially the lineshapes of the electron spectra for different NIR-XUV delays are not treated independently, but are correlated with each other by the linear chirp parameter. The quantitative evaluation of streaking spectrograms is therefore predominantly based on the TDSE-retrieval while the COE method is used to illustrate the order of magnitude of the observed time shifts. However, it should be emphasized that both techniques gave consistent quantitative results for high-quality streaking data.

The results presented so far were obtained with higher XUV photon energies than used in the first proof-of-principle experiment by Cavalieri et al. [22]. To enable a more conclusive comparison, streaking measurements on W(110) have also been performed with an XUV mirror featuring a reflectivity curve centered at ∼94 eV and a bandwidth of 6.8 eV. In these measurements, the intensity of the streaking field was deliberately reduced to minimize the ATP background in the spectral region of the W4f states which now appear at significantly lower kinetic energies. A typical streaking spectrogram obtained under these conditions is shown in Fig. 4.7 (a). In contrast to the spectrograms recorded with the higher XUV energies, almost no indication of a chirp can be detected in the streaked electron wave packets. This effect is rather attributed to the changed characteristics of the employed XUV mirror than to a photon-energy-dependent phenomenon in W(110),

Attosecond Photoemission from Clean Metal Surfaces

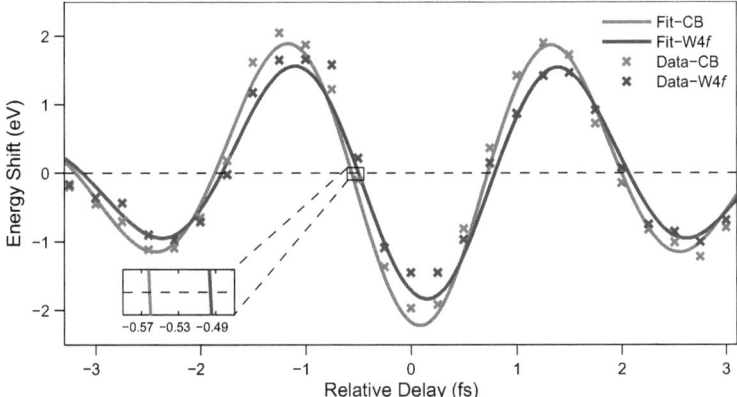

Figure 4.8: COE analysis of the streaking spectrogram depicted in Fig. 4.7 (a). The fit suggests a slightly enhanced temporal delay of $\Delta\tau \approx 60$ asec for the photoemission from the W4f states when initiated with attosecond pulses centered at $\sim$94 eV.

since this chirp was also absent in gas-phase streaking measurements performed with the same multilayer mirror. In addition, compared to excitation with $\sim$118 − 126 eV photons, the W4f-to-CB intensity ratio is noticeably smaller. This is mainly due to the opposing trend of the corresponding atomic photo-ionization cross-sections which decrease in the photon energy range of 90 − 140 eV for the W5d levels while it increases at the same time even more drastically for the W4f states [112]. As a consequence, the relative weight of the W5$p_{3/2}$ states in the total W4f feature becomes more important. Evidence for this can be found in Fig. 4.7 (b), where a quantitative analysis of a synchrotron photoelectron spectrum recorded with $\hbar\omega = 94$ eV reveals an increased contribution of the W5$p_{3/2}$ levels of 17 %. These changes have been taken into account in the corresponding TDSE-analysis of the spectrograms by adjusting the amplitudes of the individual transitions accordingly.

Unfortunately, all the streaking measurements with the 94 eV mirror have been carried out during the early stages of the project, where the cleanliness of the crystal surface could not always be guaranteed. An unambiguous interpretation was therefore possible only for a few data sets. The spectrogram depicted in Fig. 4.7 (a) was acquired shortly after flash annealing the sample to 2400 K and should therefore be representative of clean W(110). Indeed, an XUV-only photoelectron spectrum collected from this sample is in good agreement with convoluted synchrotron data (red line in Fig. 4.7 (b)). The COE analysis of this spectrogram shown in Fig. 4.8 indicates a relative time delay of $\Delta\tau \approx 60$ asec between the release of the W4f and CB photoelectrons. This is consistent with the mean time shift of $\overline{\Delta\tau} = 55 \pm 10$ asec obtained from the TDSE-analysis of 4 streaking spectrograms collected from samples exhibiting a similar surface purity. In this context, it should be mentioned that measurements conducted only under high-vacuum conditions ($p \approx 5 \cdot 10^{-8}$ mbar) yielded distinctly larger time delays of $\Delta \tau = 90 - 100$ asec, which is rather close to the value of $\Delta \tau = 110$ asec derived from a single measurement by Cavalieri et al. [22]. Under these experimental conditions, also the CB photoemission of the tungsten surface undergoes a characteristic transformation, which manifests itself in a peak broadening and a shift of the conduction band center to lower binding energies (comp. gray curve in Fig. 4.7 (b)). This can be traced back to the accumulation of contamination on the W(110) surface, as will be shown in the following section.

4.1.2 Impact of Surface Contamination

The adsorption of impurity atoms is a potential nuisance in all surface science experiments since the presence of contamination, even in the low sub-monolayer regime, suffices to alter the electronic properties of a clean surface appreciably. Considering the extremely high surface sensitivity in the streaking experiments, such impurities can also be expected to influence the measured time shifts to some extent. Great care was therefore taken to identify a possible impact of surface contamination on the experimental results.

Obviously, an *in situ* characterization of the surface quality in attosecond photoemission experiments would be advantageous. Figure 4.9 (b) compares NIR field-free photoelectron emission obtained from W(110) with 4.2 eV bandwidth XUV radiation centered at ∼118 eV for different initial conditions of the surface. The asymmetric tail of the CB emission, which is characteristic for the clean tungsten surface, disappears over time and has developed into a broad symmetric feature after ∼2 h exposure to the NIR streaking field. Similar variations in line-shape are observed after dosing a freshly cleaned surface with 6 L CO (1 L = $1.33 \cdot 10^{-6}$ mbar · s). An additional comparison with a spectrum obtained from a monolayer oxygen atoms on the W(110) suggests the O$2p$-derived states with a binding energy of ∼6 eV to be responsible for the spectral broadening in the CB region [113]. Consequently, the surface cleanliness can be judged to a certain degree by considering any deviation from the asymmetric CB line-shape as an indication for the presence of surface impurities. The analysis of a number of streaking measurements performed on W(110) crystals exhibiting different degrees of surface contamination is summarized in Fig. 4.24 (b) on page 84. Apparently, the time delays between the W$4f$ and CB photoemission extracted from these, not atomically clean, surfaces are consistently larger and cover a total range of $\Delta \tau = 40 - 70$ asec. An exemplary streaking spectrogram is shown in Fig. 4.9 (a). The corresponding COE analysis presented in Fig. 4.10 corroborates the increase of the relative time delay for insufficiently clean W(110) surfaces. Interestingly, also the surface exposed to 6 L CO exhibited a significantly larger time shift of $\Delta \tau \approx 70$ asec (see Fig. 4.11).

These findings clearly underline the benefit of retaining a sufficient spectroscopic resolution in attosecond photoemission experiments. Otherwise, changes in sample condition

Attosecond Photoemission from Clean Metal Surfaces

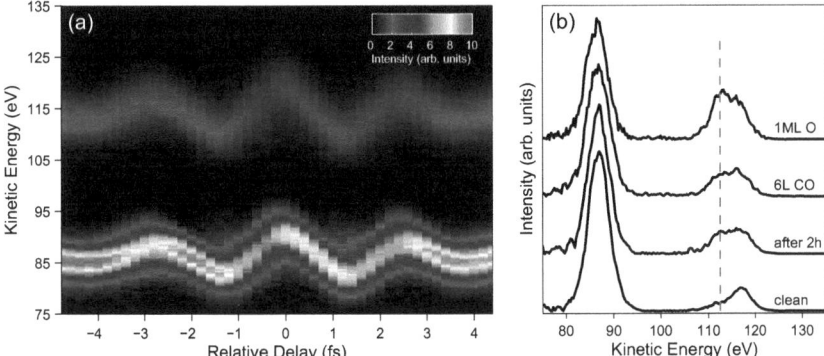

Figure 4.9: Surface contamination effects in W(110). (a) Streaked photoemission of a not perfectly clean W(110) surface. Electrons are photo-excited with sub-fs XUV pulses selected by a 4.2 eV broad multilayer reflector centered at ~118 eV and streaked with a NIR intensity of $3 \cdot 10^{11}$ W/cm². (b) Comparison of XUV-only spectra obtained from clean W(110) after different surface treatments: clean W(110), after 2 h exposure to the NIR streaking field, after exposure to 6 L CO, and after adsorption of 1 ML oxygen. The red dashed line indicates the energy position expected for O$2p$-derived states.

may remain hidden and lead to misinterpretation of the measured temporal effects. Especially in the experiments performed with the 6.8 eV broad XUV multilayer mirror, the characteristic contamination-induced change in the CB line-shape is much less pronounced and a shift of its center of gravity could only be observed for extremely contaminated samples (see Fig. 4.7 (b)). This reduced sensitivity to surface contamination entails a larger uncertainty in sample purity checks based on stationary photoemission spectra which rendered an unambiguous interpretation of many streaking spectrograms impossible.

For the sake of completeness it should be mentioned that the slight increase of the CB emission intensity for NIR-XUV delays exceeding 4 fs in the spectrogram depicted in Fig. 4.5 (a) is indicative of the accumulation of impurities on the W(110) surface in the course of the measurement. However, the impact on the time shift extracted from this particular spectrogram is negligible since the retrieved time delay remains unchanged when this part of the spectrogram is excluded in the fitting. It is also important to note that spectrograms acquired shortly after the last cleaning cycle of the W(110) crystal did not yield systematically smaller time shifts. Additional tests with significantly reduced acquisition times, resulting in a minimum of only 20 min between flash annealing of the crystal and completion of the streaking measurement, further substantiate this conclusion. A possible distortion of the distribution presented in Fig. 4.6 due to residual contamination effects is therefore rather unlikely. Strictly speaking, a possible impact of hydrogen adsorption, which is the dominant constituent of the residual gas, cannot definitely be ruled out, since hydrogen induces only minor modifications in the spectral

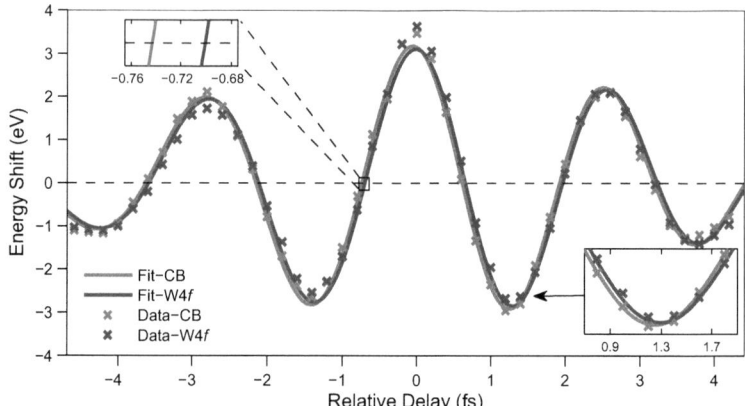

Figure 4.10: COE analysis of a streaking spectrogram obtained from a not perfectly clean W(110) surface with an XUV photon energy of ~118 eV. The fit reveals a increased time shift of $\Delta\tau \approx 40-50$ asec which has to be compared to the corresponding value $\overline{\Delta\tau} = 28$ asec characteristic of atomically clean surfaces.

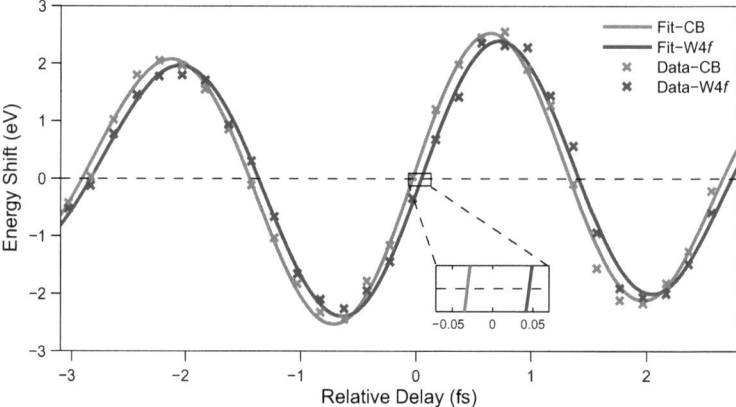

Figure 4.11: COE analysis of a streaking spectrogram obtained from a clean W(110) surface with an XUV photon energy of ~118 eV after exposure to 6 L CO. An adsorbate-induced time delay of the W4f core-level photoemission of $\Delta\tau \approx 70$ asec is clearly discernible.

shape of the conduction band. In this respect, further test experiments with saturated hydrogen monolayers on W(110) would be necessary to reach a final conclusion.

Nevertheless, the origin of the surface contamination deserves further discussion. The main constituents of the residual gas in the vacuum system at a base pressure $p \leq 10^{-10}$ mbar in the experimental area were H_2, CO, H_2O and CO_2. Considering the worst case with a sticking probability of one, the time needed for the adsorption of 1 ML atoms/molecules from the background pressure can be estimated to ~ 5 h which implies a much slower rate of contamination than observed. Probably, the local pressure in front of the surface was higher than in the remaining parts of the UHV chamber. A plausible source would be the time-of-flight spectrometer whose entrance aperture is only at a distance of 3 mm from the surface during the measurements. First tests further indicate that this contamination builds up very locally on the surface, and is restricted to an area corresponding to the spot size of the NIR focus on the sample. It was therefore not possible to detect this degradation in sample purity with standard characterization techniques, like XPS and LEED, which average over a much larger surface area. In addition, the high rate of contamination could be observed only when the surface was simultaneously illuminated with the NIR and XUV pulses whereas no sign of contamination could be found in the spectra, even after 2 h, when the NIR beam was blocked during the measurements. It is therefore conceivable, that the NIR-induced ATP electrons promote the adsorption of impurity atoms by generating radicals through dissociation of residual gas molecules in front of the crystal surface.

4.1.3 The (0001) Surface of Magnesium

The much higher reactivity of magnesium (Mg) compared to tungsten, and the time consuming cleaning procedures necessary to prepare well-ordered and atomically clean surfaces from this material [114], complicate the handling of Mg single crystals in streaking experiments. Therefore a different approach based on *in situ* deposition of Mg atoms was pursued in the experiments discussed in this section. Magnesium was sublimated at a rate of 10 ML/min onto the freshly cleaned W(110) surface from a home-built, water-cooled Knudsen-cell evaporator consisting of a resistively (AC current) heated tantalum crucible. The substrate was held at room temperature during evaporation and the crucible was thoroughly outgassed prior to each deposition process.

Figure 4.12 shows a photoelectron spectrum of ~ 30 ML Mg on W(110) recorded with $\hbar\omega = 700$ eV synchrotron radiation. The absence of emission lines associated with the C1s and O1s levels confirms the chemical purity of the deposited magnesium layers, whereas the sharpness of the hexagonal LEED pattern (inset) is indicative of epitaxial growth of bulk hcp Mg with its basal (0001) plane parallel to the W(110) surface. It is well known that these thick Mg films exhibit geometric and electronic properties indistinguishable from Mg(0001) single crystals [115]. The pristine W(110) surface is readily recovered by flash annealing the sample to 2400 K. Figure 4.13 (a) compares the photoemission from

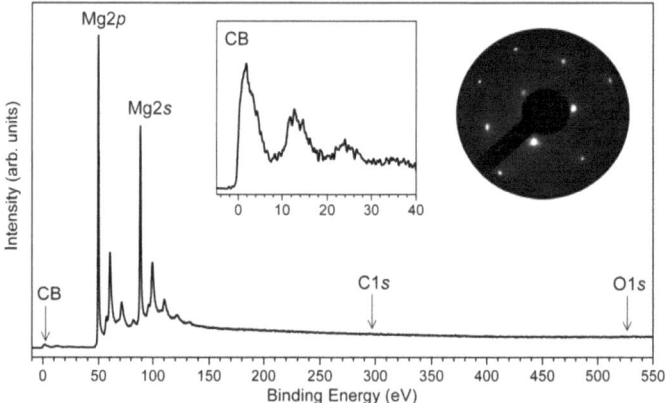

Figure 4.12: Photoelectron spectrum recorded with $\hbar\omega = 700$ eV synchrotron radiation after evaporation of $\sim$30 ML magnesium onto the clean W(110) surface at room temperature. Plasmon loss lines (mainly bulk) can be observed for all primary Mg photoemission lines. The absence of C1s and O1s photoemission lines proves the chemical integrity of the evaporated films whereas the sharp hexagonal LEED pattern (inset) is indicative of a long range ordering equivalent to Mg(0001) single crystals.

Mg(0001) with $\hbar\omega = 120$ eV synchrotron radiation (gray shaded) to the same spectrum recorded with HH radiation (circles) filtered by the $\sim$118 eV multilayer mirror. Similar to W(110), a convolution of the synchrotron data with a 4.4 eV broad Gaussian (red line) is in good agreement with the result of stationary attosecond photoemission.

In the energy range of interest for streaking experiments, the photoemission from Mg(0001) is dominated by the Mg2p core levels with a binding energy of $\sim$49.5 eV which are again composed of surface- and bulk-related electronic states. However, compared to W(110), the significantly smaller spin-orbit splitting of 280 meV of the Mg2p levels in combination with the smaller surface core level shift results in a stronger spectral overlap of the individual features. A consistent deconvolution is achieved by simultaneously fitting two Doniach-Šunjić doublets to high-resolution spectra recorded as a function of excitation energy. The result for selected photon energies is shown in Fig. 4.13 (b). Best agreement is obtained with a reasonable value for the asymmetry parameter of $\alpha = 0.13$ [12] and a surface core-level shift of 140 meV, which is in excellent agreement with an older lower-resolution photoemission study [116]. Furthermore, this quantitative analysis reveals a surface contribution of $\sim$50 % to the total Mg2p photoemission in the XUV photon energy range relevant for the attosecond experiments presented in this section.

The strong coupling of surface (sp) and bulk (bp) plasmon excitations to the Mg2p photoemission gives rise to corresponding loss lines which can be observed at 7.5 eV and 10.5 eV higher binding energy than the primary photoelectrons. They merge into a

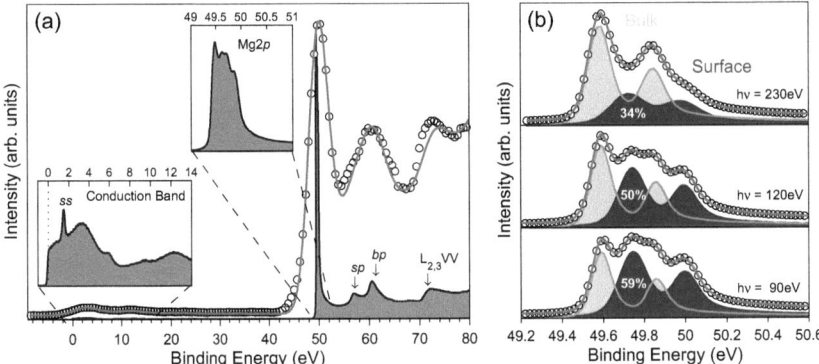

Figure 4.13: Stationary photoemission from Mg(0001). (a) Comparison of synchrotron photo-excitation at $\hbar\omega = 120$ eV (gray shaded area) and excitation with sub-fs XUV pulses centered at $\sim$118 eV (circles). Highlighted spectral features include: Γ-surface state (ss), bulk (bp) and surface plasmon (sp) satellites and the $L_{2,3}$VV Auger transition. A convolution of the synchrotron spectrum with a 4.4 eV FWHM Gaussian is shown as solid line. (b) Examples for the decomposition of the Mg$2p$ emission into bulk and surface contributions for different excitation energies.

single peak centered at $\sim$60 eV binding energy for broadband XUV excitation. The temporal characteristics of these plasmon loss electrons will be studied in more detail in Section 4.1.5 while the emphasis here will be on the primary Mg$2p$ and CB photoelectrons. Features corresponding to the excitations of multiple plasmons do also exist, but are obscured for $\hbar\omega = 120$ eV by the $L_{2,3}$VV Auger electrons stemming from the decay of the Mg$2p^{-1}$ core holes.

As expected from the small atomic photo-ionization cross-section of the Mg$3s$ levels [112] and the low density of states near E_F for sp-metals, the intensity of the CB photoemission is rather small. It comprises mainly bulk transitions for the binding energies < 6 eV [114, 117]. In analogy to the Mg$2p$ photoemission, the weaker features in the binding energy range of 7 − 15 eV can be ascribed to satellites produced by surface and bulk plasmon excitations accompanying the main CB transitions [118]. These plasmon loss lines are even more pronounced when higher photon energies are used for excitation (comp. inset in Fig. 4.12). Finally, the sharp peak at 1.6 eV binding energy is the well-known Γ-surface state (ss) existing in the gap of the surface-projected bulk band structure of Mg(0001) [114]. However, its relative intensity is obviously negligible when excited with $\hbar\omega = 120$ eV photons.

No complications due to enhanced ATP or space-charge were encountered in the streaking experiments despite the lower work function of magnesium ($\phi = 3.66$ eV) compared to tungsten. This can be regarded as another indication of the high quality of the evaporated magnesium film since surface roughness and defects are known to favor multi-photon

Attosecond Photoemission from Clean Metal Surfaces

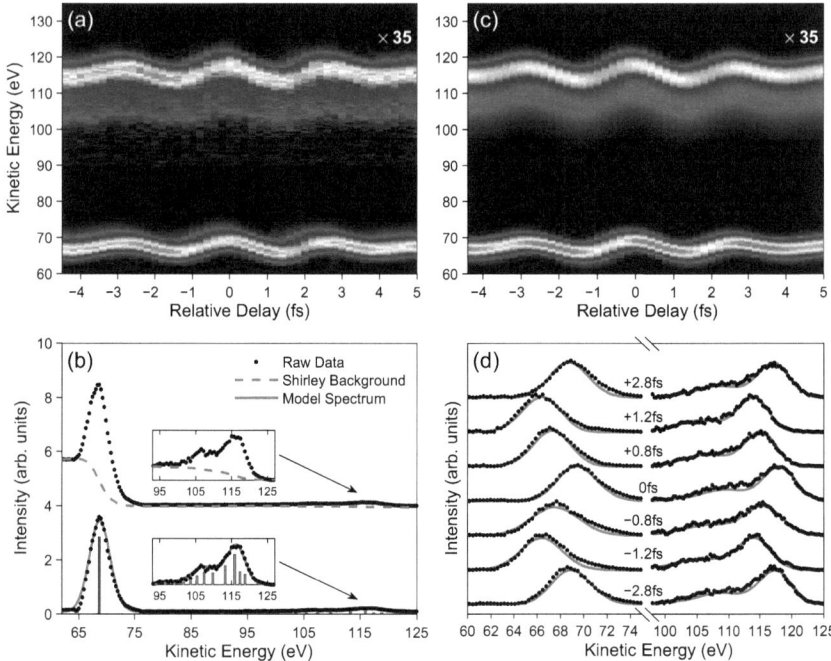

Figure 4.14: Attosecond time-resolved photoemission from the clean Mg(0001) surface. Isolated sub-fs XUV pulses centered at ∼118 eV are filtered with a 4.2 eV broad multilayer bandpass from the HH cut-off radiation and photo-excite electrons from the Mg2p and the CB states in the presence of a $1 \cdot 10^{11}$ W/cm^2 strong NIR streaking field. (a) Measured streaking spectrogram compiled from 40 individual spectra, each integrated over ∼$1.4 \cdot 10^5$ laser shots. The CB region is scaled by ×35 to visualize the good S/N ratio. (b) Top: example of the applied background subtraction scheme. Bottom: initial-state parameterization used in the TDSE-fitting of the Mg(0001) spectrograms. Insets highlight the CB region. (c) Full spectrogram reconstructed by the TDSE-retrieval algorithm. (d) Comparison of reconstructed and measured photoelectron spectra for selected relative NIR-XUV delays. The CB and Mg2p photoemission spectra are presented on separate intensity scales.

emission from solids [119]. Nevertheless, for the streaking experiments on Mg(0001), the applied NIR intensities were deliberately reduced to avoid spectral overlap between the ATP background electrons with the primary photoemission from the Mg2p core states. In this way, the same background subtraction scheme as used for W(110) could be applied (see Fig. 4.14 (b)).

The low CB photoemission intensity from Mg(0001) render streaking experiments for this system difficult. In general, longer integration times are necessary to achieve a signal-to-noise (S/N) ratio comparable to the W(110) experiments. On the other hand, the extreme

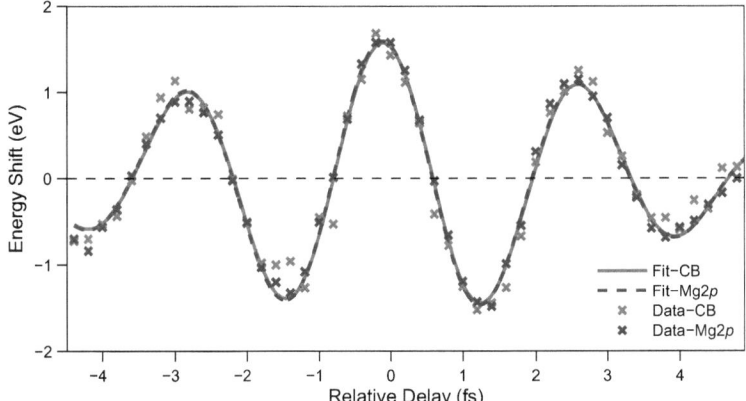

Figure 4.15: Representative COE analysis of a Mg(0001) streaking spectrogram. The first moments are calculated in the energy interval $125 - 100$ eV (CB) and $75 - 60$ eV (Mg2p) of the streaked electron distribution, respectively. The vanishing shift between the fitted streaking traces indicates a synchronous emission of the conduction band and Mg2p core electrons from the Mg(0001) surface.

reactivity of magnesium towards oxygen-containing molecules in the residual gas dictates shorter acquisition times to avoid contamination effects. This was eventually solved by collecting several spectrograms with moderate S/N ratio and renewing the Mg(0001) film after each measurement. An exemplary streaking spectrogram is depicted in Fig. 4.14 (a). The streaking effect is clearly visible for both the weak CB emission and the Mg2p core-level electrons. In particular, no build-up of contamination during the measurement (total duration ~35 min) is discernible in the CB region. Despite the low count rate in the CB region, the S/N ratio is sufficient to shed light on a possible time delay between the release of the CB and Mg2p electrons from the Mg(0001) surface. The corresponding evaluation of the spectrogram based on the COE approach is shown in Fig. 4.15. In contrast to W(110), this analysis suggests almost perfect synchronism between conduction band and core-level photoemission.

In order to further substantiate this result, the streaking spectrograms were also analyzed with the more advanced TDSE-retrieval. The parameters in the fitting routine were adjusted to provide a realistic description of the initial electronic states of Mg. The Mg2p emission is represented by a single wave packet ignoring both the spin-orbit and the surface-to-bulk splitting since these energy spacings are less than 4 % compared to the bandwidth of the exciting XUV pulses. The conduction band wave packet is phenomenologically represented by a superposition of 12 temporally synchronized transitions. Their relative strengths and central energies are indicated as red vertical bars in Fig. 4.14 (b), and are chosen to give the best agreement with a measured XUV-only electron spectrum.

Within this description, the final CB wave packet also comprises the plasmon loss satellites. A separate temporal treatment of these minor contributions in the fitting scheme led to severe convergence problems and was therefore not further pursued. The complete reconstruction of the measured spectrogram based on this initial-state parameterization is depicted in Fig. 4.14 (c). The good quality of the fit can also be judged from Fig. 4.14 (d), where measured photoelectron spectra are explicitly contrasted with the corresponding fitted spectra for selected relative NIR-XUV delays. Especially, the pronounced broadening and narrowing, observed in spectra recorded at adjacent zero-crossing of the NIR vector potential, is satisfactorily reproduced by the fit algorithm (see e.g. lineouts for -0.8 fs and 0.8 fs in Fig. 4.14 (d)).

A quantitative evaluation of 19 streaking measurements with the TDSE-retrieval gives a mean time delay of the Mg$2p$ emission of only $\overline{\Delta\tau} = 5$ asec with a standard deviation of $\sigma = \pm 20$ asec. Within the error margin, this value is still compatible with a synchronized photoemission from the magnesium conduction band and Mg$2p$ core states and therefore corroborates the result obtained from the COE analysis. The corresponding mean values for the wave packet chirp and duration retrieved by the TDSE-fitting are $\bar{b}_x = -3.1 \pm 1.6$ fs^{-2} and $\bar{\tau}_x = 386 \pm 71$ asec, respectively. Whereas $\bar{b}_x$ is almost identical to the value extracted from the W(110) measurements, the duration of the electron wave packets released from the Mg(0001) surfaces is notably larger than the average duration $\bar{\tau}_x = 280 \pm 33$ asec extracted from attosecond streaking on W(110). Since the identical XUV mirror was used in both experiments, it is tempting to relate this temporal broadening of the wave packets to a material-induced effect. Indeed, the mean electron escape depth in magnesium is larger than in tungsten (see Fig. 4.16 (a)). Thus, in Mg(0001), the electron emission from deeper atomic layers can contribute to the primary photoemission signal which broadens the temporal distribution of the emerging electrons.

4.1.4 Discussion

The results presented in the previous sections demonstrate that time delays between the photoemission of conduction band (CB) and core-level electrons from *clean* metal surfaces are significantly smaller than reported in the first proof of principle experiment by Cavalieri *et al* [22] in 2007. Obviously, a profound interpretation of such small temporal shifts is rather complicated since various, possibly compensating mechanisms may contribute to this subtle effect. Even the accuracy of ±10 asec currently achievable in state-of-the-art surface streaking spectroscopy might still not be sufficient for making reliable quantitative comparisons between different systems. The poor understanding of the exact variation of the NIR streaking field, in both its intensity and polarization, over the topmost atomic layers of a solid further adds to the complexity of the problem.

Nevertheless, it is instructive to estimate the expected order of magnitudes of the time shifts based on a simple free-electron propagation picture and assuming perfect screening of the streaking field inside the metal. Within this model, the measured time shifts

Level	E_{kin} [eV]	v_{free} [Å/asec]	λ [Å]	λ^* [Å]	τ_R [asec]	τ_R^* [asec]
Mg-CB	~115	$6.5 \cdot 10^{-2}$	4.9	5.9	76	91
Mg$2p$	67.5	$5.1 \cdot 10^{-2}$	3.7	4.5	74	89
W-CB	~115	$6.5 \cdot 10^{-2}$	3.6	4.5	55	69
W$4f$	85.5	$5.7 \cdot 10^{-2}$	3.9	4.4	69	77

Table 4.1: Basic parameters for the interpretation of streaking time shifts in attosecond photoemission from tungsten and magnesium with ~118 eV XUV pulses: λ inelastic mean free path calculated from the surface-to-bulk ratio extracted from synchrotron core-level photoemission; v_{free} velocities for free electron dispersion; τ_R resulting average travel times of the electrons to the surfaces. (*) Values derived from calculations by Tanuma et al. [121].

are identical to the difference in the average run-times τ_R of the CB and the core-level electrons to the surface (comp. Section 2.2.2). These values are determined by the inelastic mean free path of the electrons λ and the velocity of the electron inside the the solid with free electron-like dispersion $v_{free} = \sqrt{2(E_{kin} + V_0 - \phi)/m_e}$, where the kinetic energy of the electrons E_{kin} is measured with respect to the Fermi level (comp. Section 2.2.2). The inner potential V_0 is usually derived by comparing LEED I-V measurements with band structure calculation [120]. Empirically it is found that V_0 is of the order of ~10 eV for most metals and depends only slightly on the electron energy [120]. Therefore a constant value of $V_0 = 10$ eV is used throughout this thesis to calculate v_{free} in magnesium and tungsten.

Unfortunately, the inelastic mean free path λ, which is a key parameter not only in this simple model, is not known precisely enough in the electron energy range relevant for the streaking experiments. In practice, values for λ are taken from theoretical estimates relying on the extrapolation of measured optical data. However, the uncertainty associated with these calculations is rather large for low kinetic energies [121]. Experimentally, λ may be inferred from high-resolution core-level photoemission data as shown in Fig. 4.1 (b) and Fig. 4.13 (b). In a simple layer attenuation model [116, 122], the measured surface-to-bulk ratio I_s/I_b and the interlayer spacing in the direction of electron detection d are related to λ according to:

$$\lambda = \frac{d}{\ln(1 + I_s/I_b)} \quad (4.7)$$

where the relevant interlayer spacing is $d_{[110]} = 2.23$ Å for W(110) and $d_{[0001]} = 2.61$ Å for Mg(0001). Resulting values for λ, which are relevant to the streaking experiments with 118 eV XUV excitation, are compared in Tab. 4.1 to most recent calculations by Tanuma et al. [121]. Obviously, the measured values of λ are systematically smaller by ~1 Å than predicted by theory[4]. While the absolute electron run-times are notably affected by this discrepancy, the corresponding impact on the relative time shift is much less pronounced. Apparently, the average travel times of the core-level and CB electrons are almost identical for Mg(0001) which results in a vanishing time delay between their escape from the solid

---
[4] A similar deviation of 1 Å between experiment and theory was recently reported for copper [123].

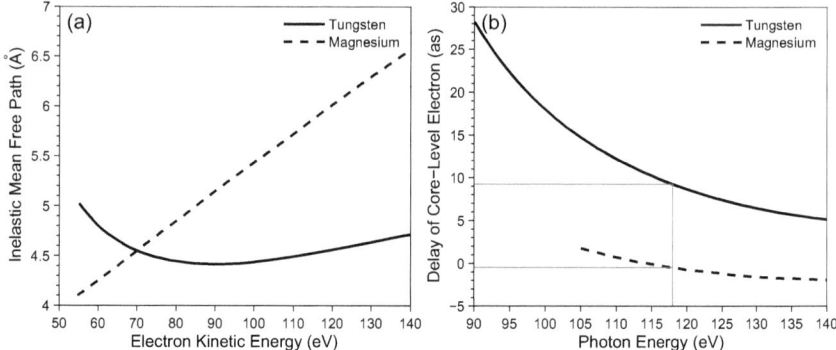

Figure 4.16: Influence of the XUV photon energy in the free-electron-propagation model. (a) Inelastic mean free path λ of electrons in bulk magnesium and tungsten according to Tanuma et al. [121]. (b) Delay in the emission of the core-level electrons with respect to the conduction band electrons calculated from their average travel times to the surface $\tau_R = \lambda/v_{free}$, with the velocity of free electrons v_{free}. Time shifts expected for $\hbar\omega_x \approx 118$ eV excitation are highlighted.

which is in reasonable agreement with the experimental result of $\overline{\Delta\tau} = 5 \pm 20$ asec obtained for this system. For W(110), this simple model predicts a delayed emission of the W4f electrons of 8−14 asec, which is at least close to the lower bound of the measured time delay of $\overline{\Delta\tau} = 29 \pm 14$ asec.

The good agreement for Mg(0001) is not surprising since this system complies best with the two main approximations underlying this simple propagation model. Firstly, the NIR streaking field can be expected to be more efficiently screened in the free-electron-like metal. This is because the frequency of the NIR field is far below the well-defined plasma frequency of magnesium which dominates its optical properties. Therefore, almost complete screening of the streaking field at the first atomic layer can be expected for this system [124]. Indeed, the refractive index of magnesium is smaller than unity in the NIR frequency range [125], which implies total reflection of the NIR pulses at the Mg(0001) surface for the grazing incidence conditions employed in the streaking experiment. This is in contrast to tungsten where the dielectric function exhibits a more complex behavior in the NIR region due to interband transitions [100]. Secondly, the free-electron-like dispersion for final-state energies 30 eV above E_F, where the unoccupied 3d states do no longer contribute to the energy bands, is well-established by synchrotron photoemission studies on Mg(0001) single crystals [114]. Thus, the small time delay can be seen as a general consequence of the monotonically increasing inelastic mean free in magnesium for electron energies > 50 eV (see Fig. 4.16 (a)).

For W(110) on the other hand, the agreement with the free-electron propagation model might be considered unsatisfactory. However, this has to be qualified to some extent because already a realistic uncertainty of ±0.5 Å [123] for the absolute values of λ translates into a corresponding uncertainty of the electron travel times of $\approx \pm 9$ asec. It is therefore difficult to decide whether the small deviation from free-electron propagation can be attributed to a modification of the photoelectron velocity by the periodicity of the crystal lattice potential according to $v_g(\mathbf{k}) = \hbar^{-1}\partial E(\mathbf{k})/\partial \mathbf{k}$ as originally suggested in [22]. Here v_g corresponds to the group velocity of the Bloch wave associated with an electron in the unoccupied energy band state $E(\mathbf{k})$.

In this context, the larger time shift of $\Delta\tau = 55 \pm 10$ asec measured for W(110) with the lower XUV excitation energy of $\sim$94 eV might be indicative of such final-state effects. Indeed, recent band structure calculations indicate that excitation with $\sim$90 eV photons promotes the W4f electrons into energy regions of the unoccupied part of the band structure with weak $\partial E(k_\perp)/\partial k_\perp$ dispersion, which results in a reduced group velocity for electrons escaping the crystal along the [110] direction of the tungsten crystal. In contrast, both the CB and W4f electrons are predicted to populate final states with predominantly free-electron-like dispersion when excited with photon energies near $\sim$120 eV [22, 60]. It has to be emphasized, however, that the quantification of the time shift obtained by $\sim$94 eV XUV excitation is only based on 4 measurements, and residual contamination effects cannot definitely excluded in these experiments. In the view of the typical scatter of time delays extracted from single streaking measurements (see Fig. 4.6), it is evident that a larger data base has to be established for this system before a decision on the importance of band structure effects in attosecond photoemission can be reached. This becomes even more obvious when considering that already the simple free-electron model predicts the right tendency of an increasing time delay for the W4f electrons with decreasing XUV photon excitation energy (see Fig. 4.16 (b)). Further, it has to be noted that the $\Delta\tau = 55 \pm 10$ asec time shift is rather close to the result from classical transport simulations reported by Lemell *et al.* [63].

Another important aspect that has to be taken into account when relating measured streaking time shifts to final-state effects concerns the wave vector $\mathbf{k}$ of the electrons contributing to the detected electron wave packets. The small cross-section for CB photoemission in combination with the low high-harmonic conversion efficiency impeded the acquisition of streaking spectrograms with sufficient S/N ratio for excitation energies exceeding 110 eV. With the current experimental setup, this could only be overcome by collecting photoelectrons within a larger acceptance angle of $\Delta\vartheta \approx \pm 20°$, enabled by an electrostatic lens incorporated in the TOF spectrometer (see Appendix A). However, this collection angle also entails an effective averaging $\Delta k_\parallel$ over the parallel component of the electron wave vectors $k_\parallel$ according to:

$$\Delta k_\parallel [\text{Å}^{-1}] = 0.512 \cos\vartheta \sqrt{E_{kin}[\text{eV}]}\, \Delta\vartheta \qquad (4.8)$$

For photoelectrons in the relevant kinetic energy range of $70 - 120$ eV, this amounts to $\Delta k_\parallel = 1.5 - 1.9$ Å^{-1}, which has to be compared to the dimension of the W(110) and Mg(0001) surface Brillouin zone of $\sim$1.5 Å^{-1} and $\sim$1.3 Å^{-1}, respectively. Thus, in principle transitions within the whole Brillouin zone can contribute to the photoemission current in the streaking measurements performed with ~118 eV and ~126 eV XUV photons. In contrast, only transitions near the Γ-point have to be considered for ~90 eV excitation, since the HH flux in this range is sufficient to perform measurements without using the electrostatic lens system of the spectrometer [5]. For a verification of band structure effects in streaking experiments, fully **k**-resolved band structure calculation, e.g. based on the inverse-LEED formalism, are required from which the mean velocity of the released CB and W4f electron wave packets can be derived by adequate averaging of the individual final-state group velocities $v_g(\mathbf{k})$ over both $\Delta k_\parallel$ and the bandwidth of the exciting XUV pulse. Further, bulk band structure calculations might not be sufficient since the dynamics probed in the current attosecond streaking experiments mainly involve the electrons in the first three atomic layers of the solid where the electronic structure can still be different from the bulk. Surface effects in photoemission must therefore be included in such calculations to achieve quantitative and predictive power. Unfortunately, these calculations are currently not available for the relevant range of final-state energies in W(110) and Mg(0001).

Explanations for the origin of time shifts in streaking experiments not relying on band-structure-dominated electron transport were proposed by Zhang *et al.* [61] and Kazansky *et al.* [59]. Although these two theoretical studies claim rather different mechanisms to be responsible for the time delay (comp. Section 2.2.2), they both agree in the conclusion that the initial-state localization is the decisive factor governing the temporal evolution of the electron wave packets. In their models, the localization of the core-level electrons is the reason for their delayed emission with respect to the CB electrons which are assumed to be completely delocalized. While this might be a rather good approximation for the conduction band electrons in magnesium, it is maybe less applicable to the electrons in the conduction band of tungsten. Here, the small dispersion of the W5d-derived electron bands implies a considerable localization of the electrons in the proximity of the ions forming the crystal lattice [101]. Following the line of arguments in [61, 59], one might expect the time delay to increase with the contrast in the spatial localization of the main initial electronic states probed in the streaking experiment. This, however, is at variance with the experimental finding of a smaller time delay in Mg(0001) than in W(110). It might therefore be concluded that the degree of localization of the initial electronic state plays only a minor role in time-resolved attosecond photoemission from single crystalline metals. In this context it should be noted that the surface state of Mg(0001) has only a very low intensity and is delocalized across the first two layers of the Mg lattice [126]. Thus, the spatial extension of the associated electron density perpendicular to the surface is comparable with the inelastic mean free path and decays rather slowly into the interior of the crystal [127]. Consequently, it does not compromise the delocalized character of the Mg(0001) conduction band feature probed in the streaking experiments.

Very recently time delays in the range of $20 - 100$ asec have been reported for photoelectrons released from different orbitals of atoms in the gas phase (see also Section 4.4.4)

---

[5]The regular acceptance angle of the TOF is $\pm 2°$ which corresponds only to $\Delta k_\parallel = 0.14 - 0.16$ Å^{-1}.

[44, 128]. It might therefore be quite possible that such atomic effects account for at least a fraction of the time shifts observed in solid-state photoemission experiments. This may have particular implication for the interpretation of W(110) streaking experiments with XUV photon energies near 90 eV, where the 17% contribution of the $W5p_{3/2}$ emission to the broad $W4f$ feature is no longer negligible. A possible intrinsic time delay between the electron release from the $W5p_{3/2}$ and the $W4f_{7/2}/W4f_{5/2}$ core-level states would inevitably affect the measured streaking time shift of their joint electron distribution with respect to the conduction band electrons. To experimentally disentangle such atomic effects from condensed matter specific contributions it would be instructive to compare streaking time shifts obtained from a solid target to the time shifts for the same element in the gas phase. The first attempt in this direction will be discussed in Section 4.4.4 for the specific case of xenon. A corresponding experiment for tungsten would be very challenging because of its extremely low vapor pressure even at high temperatures. However, valuable insights may already be gained from streaking experiments performed on a sub-monolayer of tungsten atoms supported on a substrate with a comparably weak and structureless electron emission for $\hbar\omega = 90 - 120$ eV excitation. Graphite, for example, would be a good candidate due to its low photoelectron yield and high melting point [129].

4.1.5 Electron-Plasmon Interactions

The excitation of plasmons is a clear manifestation of the many-body interactions accompanying the photoemission process in solids [130, 12]. In a classical picture, these excitations arise from the collective oscillation of the conduction band electrons against the positively charged ions of the crystal lattice. A quantum-mechanical description leads to a quantized energy spectrum of these plasma oscillations, with the plasmon being the associated energy quantum defined by:

$$\hbar\omega_p = \hbar\sqrt{\frac{n_e e^2}{m^* \epsilon_0}}. \qquad (4.9)$$

Here n_e denotes the conduction band electron density and m^* the effective mass of the electrons in the conduction band. The plasmon frequency ω_p is a decisive factor governing both the optical and electronic properties of a material. Consequently, it also plays a key role in the dynamical screening of charge perturbations in solid-state systems [17].

The sudden creation of a positively charged vacancy upon photo-excitation entails a rearrangement of the remaining electrons to minimize the total energy of the system. Recent calculations reveal that the temporal evolution of this response proceeds in two distinctly different steps. Immediately after the creation of the photo-hole, the electrons respond independently from each other to this charge disturbance, whereas for longer times, this response becomes a collective phenomenon leading to the formation of well-defined plasma oscillations [17]. The characteristic time scale for the transition between these two regimes is determined by the plasma oscillation period $T_p = 2\pi/\omega_p$. In semiconductors, this transition evolves on a femtosecond time scale and could therefore be followed in real-time

Attosecond Photoemission from Clean Metal Surfaces

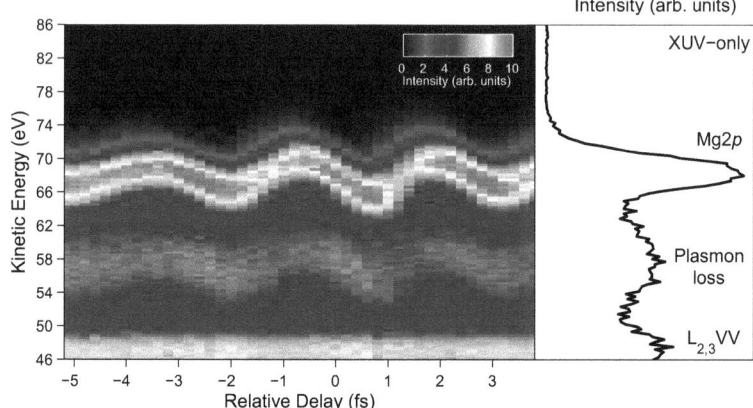

Figure 4.17: Streaking spectrogram from Mg(0001) highlighting the Mg2p emission and its first plasmon loss. High-harmonic radiation filtered by the 118 eV @ 4.2 eV bandwidth multilayer is used for excitation. No background has been subtracted. A NIR field-free spectrum is shown on the right hand side.

within a more conventional laser-based pump-probe scheme [131]. In metals, on the other hand, these dynamics are expected to proceed in the attosecond domain due to the much higher conduction band electron density. In this context, the question arises whether this ultrafast build-up of collective excitations can be probed with attosecond streaking spectroscopy.

Access to these dynamics may be provided by the pronounced plasmon loss satellites observed in photoemission from Mg(0001) (see Fig. 4.12 and 4.13 (a)). For this metal, the bulk plasmon energy of 10.5 eV implies a characteristic time scale of $T_p \approx 400$ asec for the formation of these quasi-particles. A streaking spectrogram comprising both the primary Mg2p photoemission and the first plasmon loss peak is shown in Fig. 4.17. The first plasmon loss corresponds to primary Mg2p photoelectrons which have excited exactly one bulk or surface plasmon upon their escape from the Mg(0001) crystal. This intense plasmon loss makes up more than 35 % of the total Mg2p emission strength [132]. As can be seen from Fig. 4.17, the streaking effect is clearly resolved for this loss feature, which proves that the predicted ~400 asec formation time of the plasmon excitation does not lead to a noticeable temporal broadening of the involved Mg2p electron wave packets.

Nevertheless, it might be possible that the interaction of the Mg2p photoelectrons with the plasmonic field gives rise to a temporal delay in their emission compared to Mg2p electrons leaving the solid without having excited plasma oscillations. Indeed, the COE analysis in Fig. 4.18 suggests a delayed emission of $\Delta\tau \approx 60$ asec for Mg2p electrons that have coupled to plasmon excitations during their release from the solid. A preliminary analysis with the TDSE-retrieval has also been performed. Here, the wave packet describing the plasmon loss feature was modeled by two individual electron wave packets to account for surface and bulk plasmon excitations. Their relative intensities and energy positions are extracted from the high-resolution synchrotron photoemission spectrum depicted in Fig. 4.13 (a). The time shifts retrieved from two independent measurements in this way are ∼80 asec and ∼100 asec, respectively, with only negligible influence of the applied background subtraction scheme.

At first glance, it appears that this time delay cannot be interpreted within a simple transport picture, since free-electron-like propagation in combination with the almost identical mean free paths for electrons with kinetic energies of ∼68 eV and ∼58 eV would result in a vanishing run-time difference. However, it should be emphasized that the average escape depth for electrons that are allowed to suffer one plasmon loss before leaving the solid can be larger than the escape depth of the primary Mg2p photoelectrons. This translates into increased travel times for the electrons contained in the plasmon loss peak. Assuming free electron-like velocities for the propagating photoelectrons, a relative difference in the escape depths of one Mg(0001) interlayer spacing (2.61 Å) is already sufficient to account for the relative time delay of 60 asec. Such an increase in the escape depth of one atomic layer is not unrealistic, considering that mainly losses due to plasmon excitation are responsible for the small electron mean free path in the energy range relevant for the streaking experiments discussed here [133].

Undoubtedly, further measurements on this system are necessary to provide a more reliable basis for quantitative discussions. Experimentally, these measurements are more challenging than the streaking experiments presented in Section 4.1.3, where the focus was on the relative time delay between the conduction band and Mg2p electrons. This is because the plasmon loss peak resides on a huge background of inelastically scattered electrons which may introduce some uncertainties in the COE analysis. Furthermore, the streaking field has to be limited to even lower intensities in order to retain a sufficient energy separation between the streaked plasmon loss line and the $L_{2,3}$VV Auger transition. The electron wave packets associated with these Auger transitions are not streaked since their durations exceed the period of the NIR streaking field[6]. Future streaking measurements focusing on the plasmon loss peak would therefore greatly benefit from the use of XUV pulses with higher photon energies. Finally, compared to the measurements presented in Section 4.1.3, a different setting of the electrostatic lens of the TOF spectrometer had to be used in order to optimize the collection efficiency for the energy region containing the plasmon loss (see Appendix A).

Also from a theoretical point of view, the interpretation of the Mg2p plasmon loss time delay is more complex. The synchrotron photoemission spectrum depicted in Fig. 4.13 (a) reveals that this feature actually consists of surface and bulk plasmon losses with a relative intensity ratio of ∼1:2 which are not resolved for the excitation with attosecond XUV pulses. A further distinction has to be drawn between "intrinsic" and "extrinsic" plasmons. Intrinsic plasmons are excited *during* the creation of the photo-hole, whereas

---

[6]The lower bound for the wave packet duration is set by the Mg2p^{-1} core-hole lifetime of ∼22 fs [134].

Attosecond Photoemission from Clean Metal Surfaces

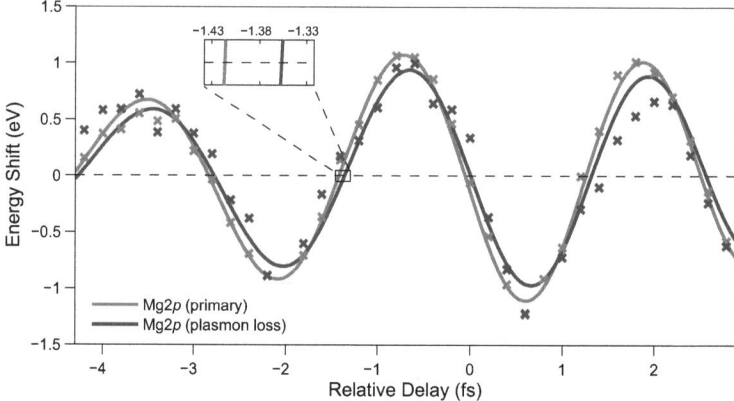

Figure 4.18: A COE analysis of the spectrogram shown in Fig. 4.17 indicates an additional time lag of $\Delta\tau \approx 60$ asec in the Mg2p electron emission when plasmon excitations couple to the photoemission process. The amplitude of the plasmon loss streaking trace is scaled by ×4. No background was subtracted from the electrons spectra.

extrinsic plasmons are excited by inelastic scattering of the photoelectron while propagating through the solid. Both excitation mechanisms contribute to the observed energy loss structure and are indistinguishable by their excitation energy.

In general, all these four energetically unresolved contributions to the plasmon satellite may exhibit different dynamics, which complicates a detailed interpretation of the measured time shift. On the other hand, an explanation solely based on electron transport arguments appears to be insufficient, since the Mg2p electrons related to intrinsic and extrinsic surface plasmon losses stem only from the first atomic layer where propagation effects should be minimal. In addition, a plasmon-induced increase of the average escape depth cannot be expected for the intrinsic part of the bulk plasmon loss feature, which amounts to $\sim 35\%$ of the total satellite intensity [12]. It is therefore still conceivable that phenomena beyond a simple electron propagation effect contribute to the measured time delay.

4.2 Effects of Chemisorption: O/W(110)

The adsorption of atoms on a surface can lead to changes in the surface electronic structure, which is often accompanied by a rearrangement of the valence charge density in the topmost layers of a solid. Depending on the strength of interaction, the adsorbate valence levels will either remain localized and almost unperturbed, or they may hybridize with band states of the substrate leading to the formation of new electronic states which can be significantly delocalized across the interface region. The presence of a chemically different species on a metal surface can therefore influence the initial-state character of the conduction band electrons which contribute to the measured photoemission current. On the other hand, the wave functions of the stronger bound core electrons in the substrate will remain unaffected by the adsorption process (besides a small shift of the core-level energies at the very surface). In particular, the spatial extent of their wave functions will hardly be perturbed. This fundamentally different behavior upon adsorption should be reflected in the time structure of electrons released from these two types of electronic states. First indications of this phenomenon have already been presented in Section 4.1.2, where both the presence of surface contamination and the adsorption of CO influenced the relative time delay between the conduction band and W4f photoelectrons. However, these systems were not sufficiently well-defined, since even CO adsorbs partly dissociatively on W(110) at room temperature which results in a rather complex structure of the adsorbate layer [135]. In the following, this adsorbate-induced effect will therefore be studied in more detail for a highly-ordered monolayer of oxygen atoms on the W(110) surface.

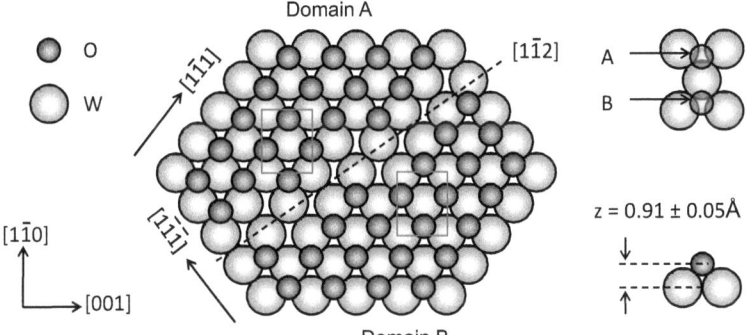

Figure 4.19: Schematic depiction of the adsorption geometry of a saturated monolayer of oxygen on W(110). Oxygen adsorbs atomically in the three-fold hollow sites of the W(110) surface. The two possible adsorption sites per surface unit cell (indicated as rectangles) give rise to the formation of two types of domains with local (1 × 1) symmetry.

Effects of Chemisorption: O/W(110)

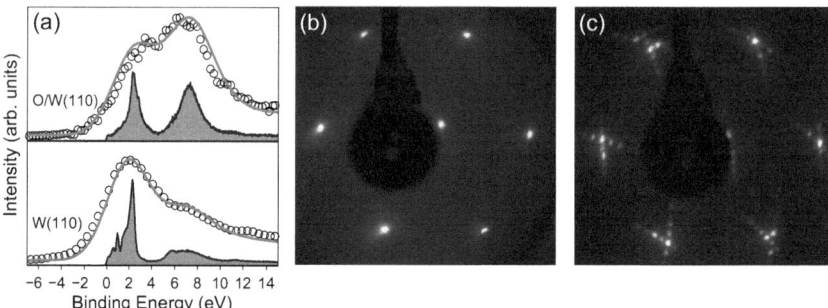

Figure 4.20: Changes in the electronic and geometric properties of W(110) upon oxygen adsorption. (a) Conduction band photoelectron spectra of W(110) recorded with $\hbar\omega = 130$ eV synchrotron radiation (gray shaded area) before and after the reaction with oxygen are compared to corresponding spectra obtained with sub-fs XUV pulses centered at $\hbar\omega\_x \approx 118$ eV (circles). Solid lines represent synchrotron spectra convoluted with a 4.4 eV FWHM Gaussian. The corresponding LEED patterns acquired with an electron beam energy of 100 eV from (b) the clean W(110) surface and (c) after completion of the O−(1 × 1)−superstructure are also shown.

4.2.1 Sample Preparation

The oxygen monolayer was prepared by either exposing the freshly cleaned W(110) surface to ∼3000 L O_2 at room temperature, or by dosing the clean surface with ∼1000 L O_2 with the sample temperature raised to ∼1000 K. Under both conditions, oxygen is known to chemisorb atomically in the three-fold hollow sites of the W(110) plane forming a densely-packed layer with (1 × 1) symmetry [136, 137, 138, 139]. Further, no bulk oxidation or adsorbate-induced reconstruction of the W(110) surface occurs during this surface reaction [140, 141]. The final structural configuration as confirmed by DFT calculations [142] and photoelectron diffraction experiments [137] is depicted in Fig. 4.19. The (110) surface of a bcc crystal offers two possible triply-coordinated adsorption sites per surface unit cell which cannot be occupied by the oxygen atoms simultaneously, therefore giving rise to the formation of two different types of domains. For high-temperature adsorption, an additional ordering of the oxygen layer can be observed which is characterized by a regular spacing of the $[\bar{1}12]$ and $[1\bar{1}2]$ domain walls in the $[1\bar{1}1]$ and $[1\bar{1}\bar{1}]$ direction [137, 138, 139]. This additional periodicity manifests itself in the corresponding LEED pattern (see Fig. 4.20 (c)) as a set of superstructure spots superimposed on the (1 × 1) pattern and trailing off in the $[1\bar{1}1]$ and $[1\bar{1}\bar{1}]$ direction. In the following, however, it will not be distinguished between these two sample preparations since the results obtained in the attosecond streaking measurements are identical. This already indicates that only the local adsorbate structure matters in these experiments.

The chemisorption of oxygen also induces pronounced changes in the surface electronic structure. This is reflected in the high-resolution conduction band photoemission spectra

Effects of Chemisorption: O/W(110)

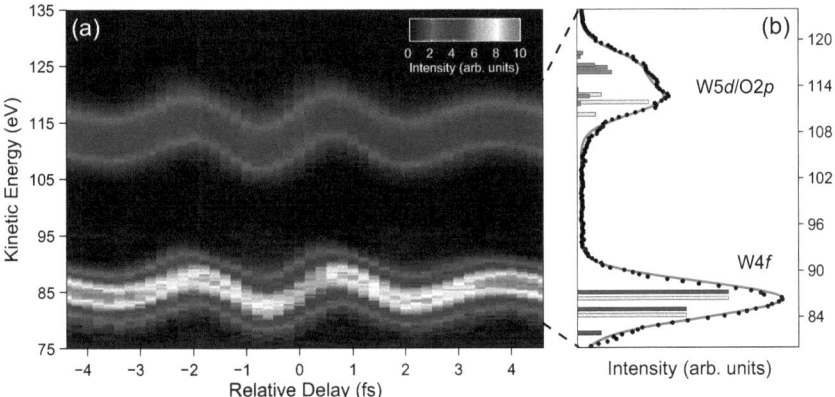

Figure 4.21: (a) Streaking spectrogram obtained from a W(110) surface covered with 1 ML oxygen. The photoelectrons are excited with sub-fs XUV pulses filtered from the HH cut-off continuum with the multilayer mirror centered near ∼118 eV, and are simultaneously dressed with a NIR field intensity of $1.8 \cdot 10^{11}$ W/cm^2. (b) A measured XUV-only spectrum of 1 ML O/W(110) (dots) is compared to a XUV-only spectrum reconstructed by the TDSE-retrieval (solid line) based on the initial-state configuration indicated as vertical bars. The light gray colored bars denote oxygen-related transitions which have been introduced in addition to the initial states of the clean surface.

depicted in Fig. 4.20 (a). New electronic states derived from the atomic O2p orbitals appear in the energy range of $5-8$ eV below E_F, but also the W5d-related energy levels are notably affected by the interaction with the adsorbate. The latter points to a strong O2p–W5d hybridization at the O/W(110) interface [143]. For the broadband attosecond excitation (circles in Fig. 4.20 (a)), all these states merge into a single broad feature in the photoelectron spectra, having now the opposite asymmetry than the conduction band feature of the pristine surface. In addition, the high electronegativity of oxygen provokes a partial charge transfer from the topmost layers of W(110) towards the adsorbate. This modified charge distribution entails an energetic shift of the surface W4f states to ∼1 eV higher binding energies compared to the clean surface [141]. In contrast to the CB emission, this change in initial states is hardly detected in the attosecond photoemission spectra, where it results only in a minor broadening of the W4f feature (comp. Fig. 4.9 (b)). The O2s-derived levels, expected at a binding energy of ∼23 eV, could not be observed in the experiments due to their low photo-ionization cross-section.

4.2.2 Streaking Experiments

Figure 4.21 (a) shows a typical streaking spectrogram obtained from a W(110) surface saturated with 1 ML oxygen. All measurements reported in this section were performed

Effects of Chemisorption: O/W(110)

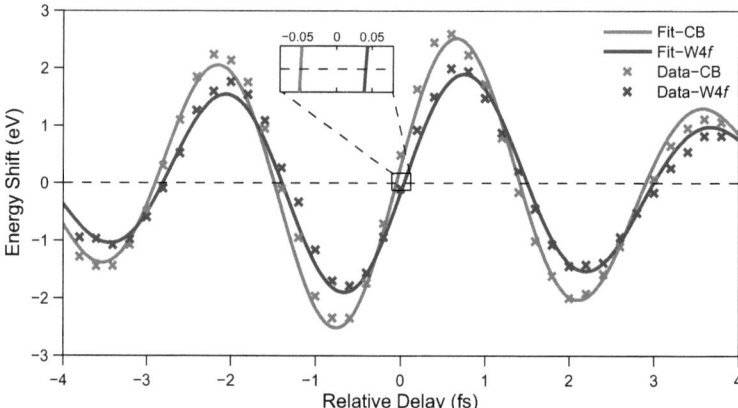

Figure 4.22: COE analysis of the streaked photoemission originating from 1 ML oxygen on W(110). Compared to the clean surface, the adsorption of oxygen increases the time shift between the W4f and the CB photoemission to $\Delta\tau \approx 80 - 90$ asec.

with the 118 eV @ 4.2 eV bandwidth XUV mirror. NIR intensities similar to those in the streaking experiments on clean W(110) could be applied without running into problems with space-charge or enhanced background of ATP electrons. This can be expected because the presence of oxygen further increases the work function of W(110). On the other hand, it may also indicate that the surface roughness is not significantly changed upon oxygen adsorption. Indeed, the distance between the oxygen atoms sitting in the threefold hollow sites and the atoms in the first plane of the substrate is less than 1 Å which results in a relatively smooth surface contour [137, 138].

The analysis of the streaking spectrogram according to the COE method is presented in Fig. 4.22. It reveals a sizable shift between the two streaking traces corresponding to a time delay of the W4f photoelectrons of $\Delta\tau \approx 80 - 90$ asec with respect to the electron release from the O2p/W5d-hybrid conduction band. This further corroborates the experimental trend suggested in Section 4.1.2, where an enhancement of the time shift was found in the presence of oxygen-containing adsorbates on the W(110) surface. A corresponding analysis of the spectrogram with the TDSE-retrieval is summarized in Fig. 4.23. Here, in accordance with [143], three additional transitions with binding energies between $5-8$ eV have been included in the description of the joint CB wave packet. Furthermore, the oxygen-shifted surface components of the W4f doublet are incorporated in the final core-electron wave packet [141]. The relative intensities of the oxygen-induced CB transitions are phenomenologically adjusted to match the XUV-only spectrum collected from the oxygen monolayer on tungsten (see Fig 4.21 (b)). The relative strength of the W4f surface component is fixed to 50 %, i.e. the same surface-to-bulk intensity ratio as measured for

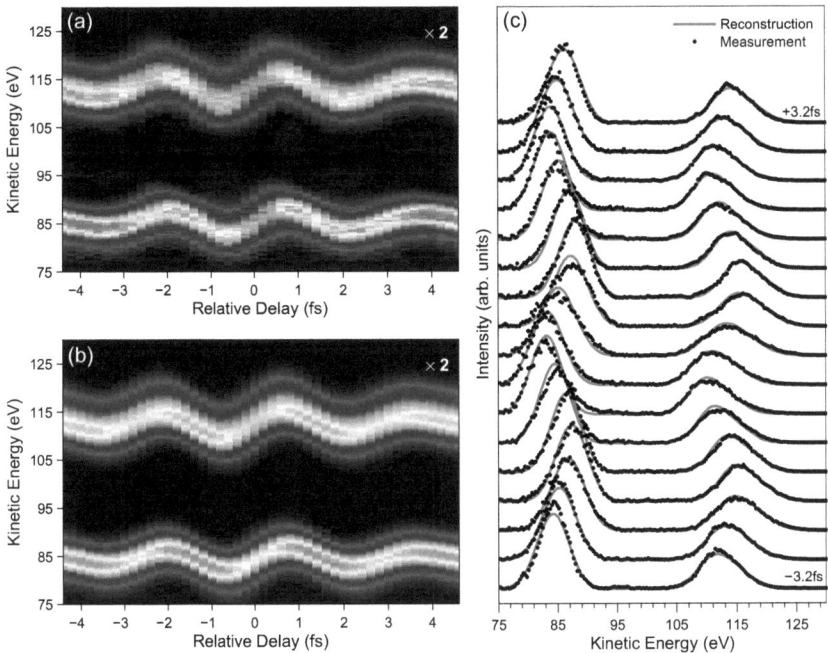

Figure 4.23: Example for the evaluation of a streaking spectrogram of 1 ML oxygen on W(110) with the TDSE-retrieval. (a) Experimental data. (b) Reconstruction of the spectrogram optimized by the TDSE-fitting routine. The CB emission is scaled by ×2. (c) Comparison of data and reconstruction at selected relative delays between the NIR and XUV pulses. See text for further explanation.

clean W(110). This is a reasonable assumption, since the ratio is primarily determined by the surface sensitivity associated with the photon energy used in the experiment.

As can been seen from the comparison of measured and reconstructed photoelectron spectra in Fig. 4.23 (c), the parametrization of the conduction band wave packet used in the TSDE-fitting procedure tracks almost perfectly the line-shape variation of the CB emission feature over the entire range of temporal overlap between the XUV and NIR pulses. It is therefore not stringently required to implement an additional fit parameter in the TDSE-routine that takes a possible relative time delay between the $O2p$- and $W5d$-dominated states into account. Quite generally, such a time shift between two spectrally overlapping features gives rise to a spectral deformation of the electron distribution as a function of NIR-XUV delay which resembles those induced by a linear chirp of a streaked wave packet (see Appendix C). However, the significant chirp already introduced by the XUV mirror itself complicates such an advanced analysis of the streaking spectrograms.

Effects of Chemisorption: O/W(110)

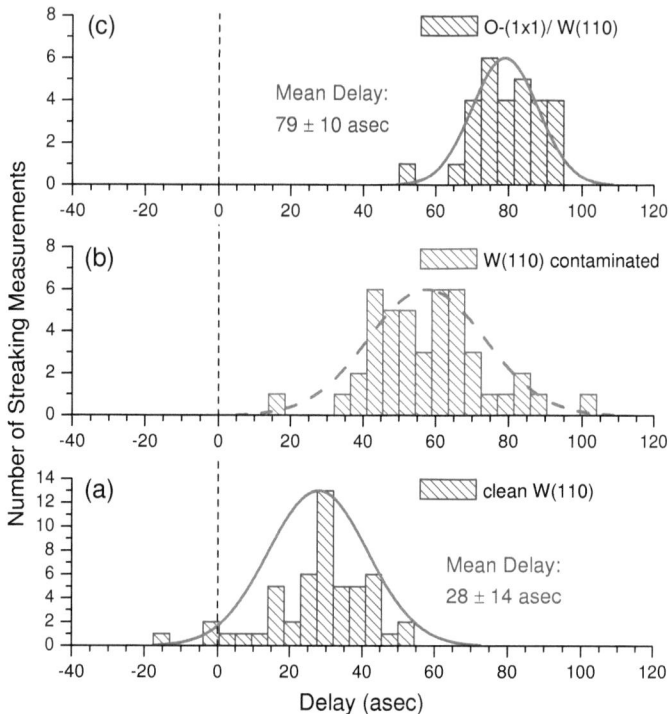

Figure 4.24: Summary and statistical analysis of the time delays measured between conduction band and W4f electrons emitted from W(110) for different initial conditions of the surface: (a) clean W(110), (b) W(110) exhibiting surface contamination and (c) W(110) saturated with one monolayer oxygen. All time delays have been extracted from the spectrograms using the TDSE-retrieval algorithm. The exciting XUV radiation was centered either at ~118 eV or ~127 eV, without any detectable influence on the observed time shifts. The classification of the surface condition in (a) and (b) is based on the analysis of the conduction band line-shape (comp. Fig. 4.9 (b)).

The statistical analysis of the time delays extracted from 29 spectrograms with the TDSE-retrieval is presented in Figure 4.24 (c). All the retrieved time shifts are distinctly larger than for clean W(110), giving a mean delay of $\overline{\Delta \tau} = 79$ asec with a standard deviation of $\sigma = \pm 10$ asec. This is compatible with the COE result and clearly establishes the oxygen-induced time delay as a robust effect beyond all experimental uncertainties. It should be remembered that the oxygen-covered W(110) surface is comparably inert at room temperature. Therefore longer integration times are possible without any disturbance by surface contamination effects. Consequently, no dependence of the time delay on the NIR exposure time was observed for this system. Nevertheless, the magnitude of

Effects of Chemisorption: O/W(110)

this adsorbate-induced time shift seems to depend sensitively on the oxygen coverage. First measurements performed on W(110) covered with only $\sim 0.75\,\mathrm{ML}^7$ oxygen revealed a slightly smaller time delay of $\overline{\Delta \tau} = 57 \pm 9$ asec (average of 5 measurements).

4.2.3 Interpretation

In general, changes in the electronic structure of a solid are inextricably linked to a corresponding change of its optical properties via the dielectric function. This has two main implications for the interpretation of attosecond streaking experiments on adsorbate-covered surfaces. Firstly, the initial as well as the final states defining the properties of the released electron wave packets can differ from the pristine case. Secondly, modifications of the NIR streaking field at the newly formed interface due to changes in refraction and screening have to be taken into account. Although the notion of screening and refraction is debatable when applied on atomic length scales, two limiting cases might nevertheless be distinguished. If the NIR screening efficiency of the adsorbate layer is much weaker than for the substrate, photoelectrons excited inside the adsorbate can be considered to experience instantaneously the full strength of streaking field, i.e. similar to the situation in gas-phase experiments. Measuring the streaking time delay between these photoelectrons and those originating from the underlying substrate would then reveal the average travel time τ_R of the substrate electrons to the substrate-adsorbate interface [144]. If on the other hand the screening efficiency of the overlayer is similar to the substrate, the apparent travel time of the substrate electrons will be increased by the time required to traverse the adsorbate layer. This effect may be slightly balanced in the relative delay measurement by the fact that now also the electrons released from the adsorbate do no longer respond immediately to the streaking field, but have to travel a certain "distance" before being exposed to the full strength of the NIR field.

Transferring these basic ideas to the system of an oxygen monolayer on W(110) is obviously more complex. Whereas the W4f electrons can be safely ascribed to originate solely from the substrate, the conduction band emission comprises both contributions from the overlayer and the substrate. Apparently, more information on the initial spatial distribution of these electrons within the interfacial region is required for interpreting the time evolution of their photoemission. Figure 4.25 shows the calculated layer-resolved density of states (DOS) for the O−(1 × 1)/W(110) system [145]. There is a striking similarity between the DOS at the oxygen adsorbate layer and the DOS in the first atomic layer of W(110). Compared to the DOS of bulk W(110), additional states with energies of $4-8$ eV below E_F are clearly resolved in *both* the adsorbate and the first atomic layer of tungsten. Moreover, in the first two tungsten layers, the structure of the originally W5d-dominated energy states in the energy range of $0-4$ eV below E_F notably differs from the bulk, and electronic states with these energies can also be observed within the oxygen overlayer. This is a clear evidence for a strong mixing between the electronic states at the inter-

---
[7]This coverage is an estimate based on the simultaneous observation of a (2 × 2) LEED pattern and XUV-only photoelectron spectra similar to those collected from O−(1 × 1)/W(110) [136, 142].

Effects of Chemisorption: O/W(110)

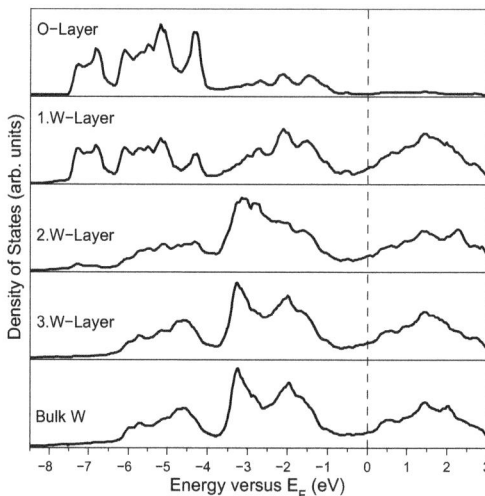

Figure 4.25:
Layer-resolved density of states (DOS) calculated for one monolayer atomically adsorbed oxygen on W(110) [145]. The DOS is projected onto the positions of the oxygen adsorbate (top trace) and the position of the tungsten atoms in the first, second and third plane of the W(110) substrate. The bottom trace represents the DOS of bulk tungsten. The enhanced DOS in the energy range of 5 − 8 eV below the Fermi level for the first layer of tungsten is indicative of a strong hybridization between the O$2p$ and W$5d$ states across at the oxygen-tungsten interface.

face. In particular, it further proves that the characteristic peaks observed at ∼6 eV and ∼2 eV binding energy in the photoemission spectra (see Fig. 4.20 (a)) cannot simply be identified with electronic states that are initially either entirely localized at the adsorbate or in the W(110) substrate, respectively. This is in agreement with the qualitative result obtained by the TDSE-analysis of the streaking spectrograms which suggests an almost identical time evolution of the electron wave packets associated with these two spectrally overlapping features.

For a first understanding of the time structure of the electrons released from this hybridized conduction band, it might therefore be sufficient to analyze the depth distribution of the total density of electrons near the Fermi level, irrespective of their exact energies. Figure 4.26 shows the valence charge density, integrated over the energy interval of 0 − 9 eV below E_F, for the first few atomic layers of clean and oxygen-covered W(110) in a false-color representation. The comparably strong localization of these electrons to the atomic positions is clearly preserved upon oxygen adsorption. Given the short inelastic mean free path of $\lambda = 4 - 5\,\text{Å}$ for the CB photoelectrons excited by the XUV pulses in the streaking experiments (see Tab. 4.1), only the variation of the average charge distribution in a volume comprising the three topmost layers is relevant for the measured photocurrent. Due to the shorter W-O bonding length of ∼1 Å compared to the W-W interlayer spacing, the center of gravity of the valence charge distribution inside the probed volume is effectively shifted closer to the solid-vacuum interface upon oxygen adsorption. Consequently, the conduction band wave packets launched from oxygen-covered W(110) arrive slightly earlier in vacuum than the corresponding wave packets released from clean W(110) surface. In view of the valence charge density depicted Fig. 4.26, and considering that screening of the NIR field in the solid is mainly a response of the valence

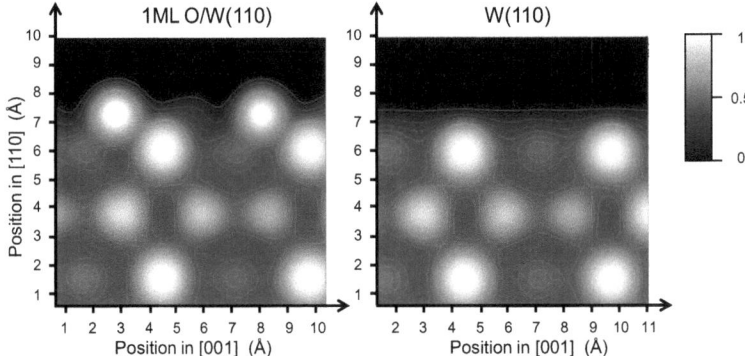

Figure 4.26: False-color representation of the integrated valence charge density in a plane perpendicular to the (110) surface for 1 ML O/W(110) (left) and clean W(110) (right) [145].

electrons, one might argue that the chemisorbed oxygen monolayer also shifts the onset of the streaking effect further away from the first atomic layer of W(110), by a distance comparable to the ∼1 Å W-O spacing in the [110] direction [137, 138]. Both effects will obviously increase the relative time delay between the W4f core-level electrons and the conduction band photoelectrons, in accordance with the experimental observation. On a quantitative level, one would expect this relative time delay to increase by at least the additional travel time of the W4f electrons through the oxygen adlayer. However, assuming free-electron-like dispersion of the 85 eV electrons, this amounts only to 18 asec and does therefore not account for the full difference of ∼50 asec between clean and oxygen-covered W(110). On the other hand, the temporal effect of the net valence charge accumulation near the solid-vacuum is difficult to quantify exactly but can be expected to be of the same order of magnitude.

In the discussion presented so far, only changes in the initial states of the O/W(110) system have been considered. However, the possibility of oxygen-induced final-state effects on the propagation of the outgoing electron wave packets cannot be definitely excluded and may contribute to the observed time shift. Finally, it should be emphasized that the absolute escape time τ_R for W4f electrons in clean W(110), which should indeed be in the range of ∼70 − 80 asec according to the free-electron-like propagation model (see Tab. 4.1), does not directly reveal itself in the experiment discussed in this section. Such a measurement would require an energy level as a reference that is strictly localized in the adsorbate layer and does not spectrally overlap with any other primary photoemission from the substrate. In contrast to oxygen, this adsorbate should not significantly perturb the charge distribution at the interface and leave the streaking field inside the overlayer unscreened. An experimental approach along these lines is presented in Section 4.4, where the focus will be on the impact of rare-gas adsorption on the streaking time shifts.

4.3 Attosecond Dynamics at Metal-Metal-Interfaces

Inelastic scattering of photoelectrons inside the solid prior to their escape into vacuum is the reason underlying the surface sensitivity of solid-state photoelectron spectroscopy. Quantified by the kinetic energy dependent inelastic mean free path, it determines the average travel time of the emerging primary photoelectrons. These transport phenomena served as a basis for the discussion of the time shifts discovered in the experiments presented so far, but a definite experimental proof for the importance of these effects is still missing. Conceivable schemes to tackle this question experimentally are sketched in Fig. 4.27. They all rely on using the thickness d of a metallic overlayer as an additional control parameter for tuning the impact of transport-related effects in streaking measurements. In the conceptually most comprehensive approach (scenario #2), the evolution of

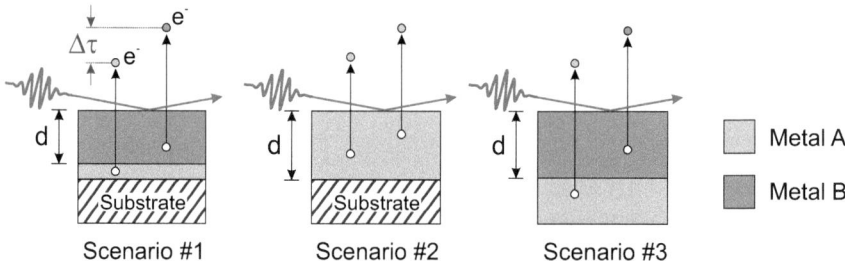

Figure 4.27: Different scenarios for the investigation of transport effects in solid state attosecond photoemission. Scenario #3 is explored in the experiments presented in this section. See text for discussion.

the relative time delay of photoelectrons released from the same metallic material is studied as a function of its thickness. By controlling the number of atomic layers, one should be able to follow the continuous transition from atomic-like to transport-dominated time shifts. The influence of the material's band structure on the electron propagation may be more conveniently disentangled within scenario #1. Here, the emission of primary photoelectrons from a buried metal monolayer is timed with respect to electrons released from a metallic overlayer consisting of a different material. The importance of band structure for the propagation of the monolayer photoelectrons might then be investigated by varying the overlayer material. However, in a realistic experiment, these metal mono-/multilayer films have to be supported by a (conducting) substrate which has to meet two mayor requirements. First of all, its photoemission signal should not interfere in the resultant electron spectrum with the line positions of interest originating from the metal adlayers. This is particularly demanding with regard to the conduction band photoemission. Secondly, the substrate must allow a nearly perfect layer-by-layer growth mode with atomically sharp interfaces between the individual adlayers, i.e. the possibility of interdiffusion, alloying or agglomeration into three-dimensional islands has to be excluded in order to facilitate interpretation. Further constraints imposed on the selection of suitable electronic states in terms of cross-sections and binding energies, due to the low HH flux at higher photon energies, render investigations according to the scenarios #1 and #2 currently impossible.

On the other hand, relaxing the restrictions on the substrate photoelectron emission characteristics significantly expands the opportunities for exploring transport phenomena within a more flexible experimental scheme depicted as scenario #3 in Fig. 4.27. In this approach, the temporal evolution of primary substrate photoelectrons traversing a capping layer of variable thickness is studied. However, a complete analysis of the time shifts occurring in such systems is obviously more complex, since they involve electron propagation inside the substrate material which cannot be controlled. Nevertheless, a cumulative propagation effect should still manifest itself in the evolution of the relative time delay between electrons released from substrate states and those excited within the overlayer when measured as a function of the overlayer thickness. Fortunately, the closely-packed (110) surface of tungsten has proven to be an ideal substrate for the growth of epitaxial layers of various metals, including magnesium [115, 146, 147, 148]. This material combination is therefore a promising starting point for probing the ultrafast motion of excited electrons in well-controlled condensed-matter systems with attosecond streaking.

4.3.1 Heteroepitaxy of Magnesium on W(110)

Ultrathin magnesium films were deposited from a Knudsen-cell-type evaporation source onto the freshly cleaned W(110) surfaces (see Section 4.1.1). The growth rate was calibrated by means of thermal desorption (TD) spectroscopy and tuned to a convenient value of $0.2 - 0.3$ ML/min. Figure 4.28 shows a series of TD spectra obtained from W(110) for different initial coverages Θ of Mg. For small coverages, a single desorption peak appears at $\sim$740 K with its maximum gradually shifting to higher temperatures as coverage increases. Upon further Mg deposition, this desorption signal finally saturates in intensity with a peak desorption temperature of $\sim$770 K. An additional desorption signal between $450 - 570$ K is observed for higher coverages which grows continuously even for prolonged Mg deposition. Consequently, the desorption peak at $\sim$740 K can be related to a W(110) surface saturated with one monolayer Mg, whereas the signal at lower temperature must be assigned to the desorption of additional Mg adlayers covering the Mg/W(110) interface [149]. In the following, all coverages will be given relative to the saturated monolayer.

Similar to the sub-monolayer regime, a continuous shift to higher temperatures with increasing coverage can also be observed for the second desorption peak in the range of $1\,\text{ML} < \Theta < 2\,\text{ML}$ (see Fig. 4.28 (c)). This behavior might be related to the attractive Mg-Mg interaction within the growing film, and/or the prevalence of nearly zero order desorption kinetics which is characteristic for desorption from 1D structures and the edges of 2D islands [25]. For initial coverages exceeding 2 ML, the multilayer TD traces reveal the presence of an additional desorption state with distinct desorption temperature, which can be associated with the desorption of atoms from the third Mg layer. The

Attosecond Dynamics at Metal-Metal-Interfaces

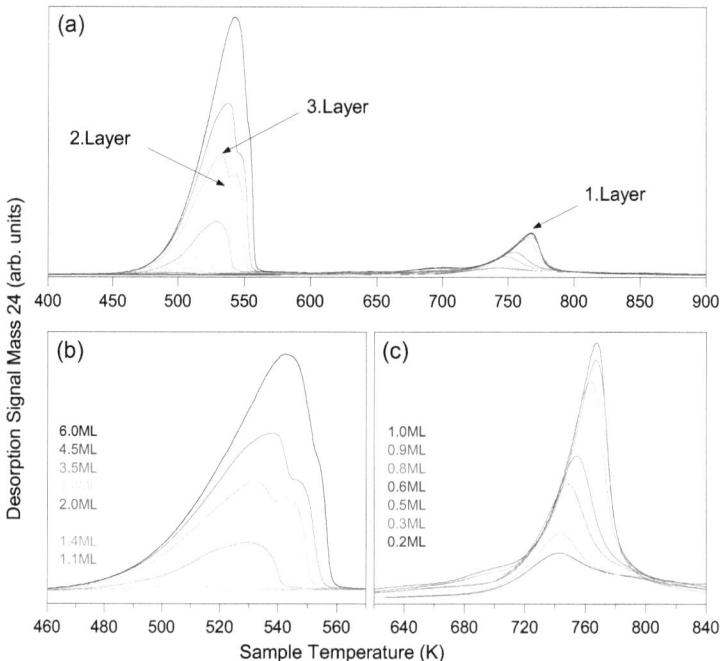

Figure 4.28: Thermal desorption from Mg/W(110). (a) Desorption spectra obtained with a temperature rate of 2 K/s for various initial magnesium coverages. (b,c) Expanded views of the temperature regions featuring multilayer and sub-monolayer desorption, respectively. Coverages in units of monolayers (ML) are derived from the area ratio of the total desorption signal and the peak area of the saturated monolayer. Up to three distinct desorption maxima are observed and can be associated with the sequential sublimation of the first three atomic layers of magnesium from the W(110) plane.

absence of any further characteristic desorption peaks for higher coverages reflects the diminishing influence of the W(110) substrate on the nucleation of the Mg adatoms as the film thickness increases. The sequential appearance of desorption states in the TD spectra related to the first, second and third Mg adlayer strongly supports that the growth of Mg on W(110) proceeds layer-by-layer even throughout this early stage of the nucleation process, whereas for other heteroepitaxial systems a Stranski-Krastanov behavior with the intermediate formation of three-dimensional islands is frequently observed [148, 25].

The exact topology of Mg layers on W(110) in the coverage regime < 10 ML has been

Attosecond Dynamics at Metal-Metal-Interfaces

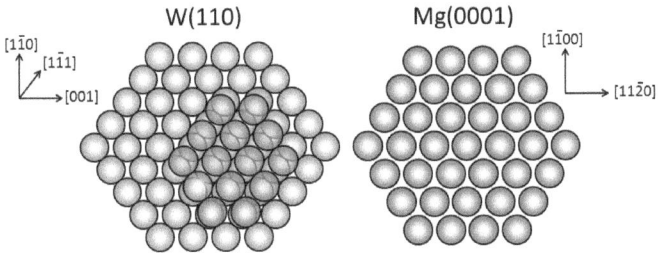

Figure 4.29: Schematic depiction of the ideal surface structures of W(110) and Mg(0001).

studied in great detail by Aballe *et al.* using LEED and LEEM[8] [148]. Their results reveal the growth of bulk-like Mg(0001) on W(110) already for $\Theta \geq 3\,\text{ML}$, with the densely-packed rows of Mg(0001) in Mg[11$\bar{2}$0] aligned parallel to the W[001] direction of the W(110) substrate (see Fig. 4.29). Below this coverage, the strongly anisotropic in-plane lattice mismatch (20 % in W[1$\bar{1}$0] and 1.5 % in W[100]) forces the Mg atoms to arrange in a slightly distorted hexagonal superstucture, which is in registry with the tungsten substrate in W[100] and has 7/8 coincidence along the W[1$\bar{1}$1] direction for the saturated monolayer. The lattice of this layer is slightly compressed compared to the ideal hexagonal Mg structure resulting in a $\sim$10 % higher atomic density compared to a (0001) plane in bulk Mg [148]. This densely-packed layer is identical to the overlayer left behind on the W(110) surface after thermal desorption of Mg mulilayers.

The evolution of the interfacial electronic structure of Mg/W(110) can be followed by high-resolution synchrotron photoemission. Figure 4.30 depicts the Mg$2p$, W$4f_{7/2}$ and the conduction band photoemission as a function of Mg coverage. The interaction of the Mg atoms with the W(110) surface becomes readily apparent when analyzing the surface-core-level-shifted component of the W$4f_{7/2}$ emission. Compared to the W$4f_{7/2}$ surface emission of clean W(110) S_W, the presence of Mg induces an additionally shifted component at $\sim$130 meV lower binding energy. This new interface feature I must therefore be related to tungsten surface atoms in direct contact with the Mg adlayer. Consistently, this peak coexists with S_W for sub-ML coverages, but is the only component persisting for $\Theta \geq 1\,\text{ML}$. Surprisingly, also the W$4f_{7/2}$ component B_W at higher binding energy undergoes a similar transformation, resulting in a new electronic state B'_W shifted to $\sim$60 meV lower binding energies. No further changes can be identified in the W$4f_{7/2}$ spectra after completion of the Mg monolayer, besides a continuous exponential damping of the total intensity with increasing thickness of the Mg overlayer due to inelastic scattering of the photoelectrons.

These changes induced by Mg adsorption in the W$4f_{7/2}$ emission might be of purely electronic nature, i.e. due to a redistribution of the valence charge driven by the hybridization

---
[8]Low Electron Energy Microscopy

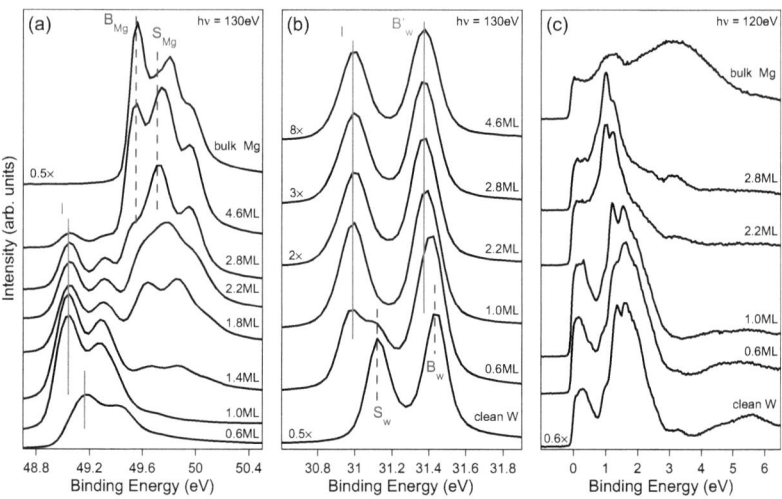

Figure 4.30: Characterization of the interfacial electronic properties of Mg layers on W(110) with synchrotron-based photoemission spectroscopy. The evolution of (a) the Mg2p, (b) the W4$f_{7/2}$ and (c) the conduction band photoemission is followed as a function of Mg coverage in normal emission (acceptance angle ≈ ±3°). Spectra collected from pure Mg(0001) and W(110) are shown as a reference. Line positions of spectral features induced by the mutual interactions of Mg and W atoms at the interface are indicated by solid lines, while electronic states characteristic of bulk Mg(0001) and W(110) are highlighted by dashed lines. See text for a detailed assignment of the individual components.

of Mg- and W-related states, or they may involve a rearrangement of atomic positions at the interface, leading to a partial intermixing or surface alloying between Mg and tungsten [150]. In this context it should be noted that no stable Mg-W bulk alloys are reported in the literature. Furthermore, no Mg-related photoemission signal could be detected from initially Mg-covered W(110) surfaces after annealing the sample to ∼850 K. This proves the complete desorption of Mg from the W(110) surface thereby ruling out the formation of a temperature-stable Mg-W surface alloy. Moreover, assuming an atomically sharp Mg-W interface, the measured attenuation of the W4f emission strength of -42 % for the monolayer-covered W(110) surface can be converted into an effective thickness of the Mg overlayer [108]. With the measured inelastic mean free path $\lambda = 4.5$ Å, an effective thickness of ∼2.5 Å is derived which is only slightly smaller than the interlayer spacing $d_{[0001]} = 2.61$ Å in bulk Mg(0001). A strong intermixing of Mg and W atoms at the interface is therefore also extremely unlikely for the temperature range < 850 K.

On the other hand, a strong hybridization between the surface resonance of W(110) and the precursor of the Mg(0001) surface state evolving in the thin Mg films was recently concluded from similar photoemission studies on the Mg/W(110) system [151, 146]. Figure 4.30 (c) shows the evolution of the conduction band photoemission for progressing Mg deposition. Up to the monolayer coverage, a strong damping of the overall intensity can be observed but only minor changes in the spectral shape. In contrast, for $\Theta > 1\,\text{ML}$, a distinct spectral narrowing occurs in the energy range of $0.5 - 2.5$ eV below E_F which can be interpreted as a gradual transition from a conduction band dominated by the W(110) surface resonance to a conduction band increasingly influenced by the Mg(0001) surface state. Nevertheless, the strong interaction at the interface implies that already the conduction band states of 1 ML Mg/W(110) carry a strongly mixed Mg-W character, even though this is not yet reflected in their spectral appearance [151, 146]. The associated rearrangement of the electronic charge at the interface seems to involve also the sub-surface region of W(110), which would be a plausible explanation for the ~60 meV shifted $W4f_{7/2}$ bulk component B'_W. Considering the high surface sensitivity for $\hbar\omega = 130$ eV excitation, B'_W can be primarily assigned to the atoms in the second layer of W(110) which are only indirectly affected by the interaction at the Mg-W interface.

The progression of the Mg$2p$ core-level emission with Mg coverage is summarized in Fig. 4.30 (a). For $\Theta < 1\,\text{ML}$, only a single doublet is observed that gains intensity and shifts to lower binding energies as more Mg atoms assemble on the surface. This continuous peak shift might be attributed to the growing size of the 2D Mg islands on the W(110) substrate, which makes the extra-atomic screening of the Mg$2p^{-1}$ core holes more efficient. As soon as the W(110) surface is completely covered by the Mg monolayer, this interface component I remains centered at a constant binding energy of 49.0 eV with its intensity being exponentially attenuated upon further Mg deposition. In the range of $1\,\text{ML} < \Theta < 2\,\text{ML}$, additional emission lines can be observed at higher binding energies that exhibit a rather complex evolution. The signal is at least composed of two Mg$2p$ doublets, indicating the existence of inequivalent binding configurations for Mg atoms on the saturated Mg-W interface. Temperature-dependent measurements combined with a detailed LEED/LEEM analysis would be necessary to explore the origin of this phenomenon. The characteristic Mg$2p$ bulk and surface component of Mg(0001) begin to develop for $\Theta > 3\,\text{ML}$, with the intensity of the bulk component B_{Mg} growing initially slower than the surface component S_{Mg}. The subsequent appearance of I, S_{Mg} and B_{Mg} is typical for the adsorption of alkaline and alkaline-earth metals on W(110) and indicates the formation of relatively sharp interfaces during the growth process [150].

In summary, it can be concluded that magnesium layers on W(110) represent sufficiently well-defined systems with atomically sharp interfaces, ideally suited for studying electron transport phenomena on the atomic length scale. However, subtle modifications and evolutions of the electronic properties at the Mg-W interface and within the magnesium overlayers do nevertheless exist. They might become important for a detailed interpretation of the streaking experiments performed on this system, which will be presented in the following subsection.

Attosecond Dynamics at Metal-Metal-Interfaces

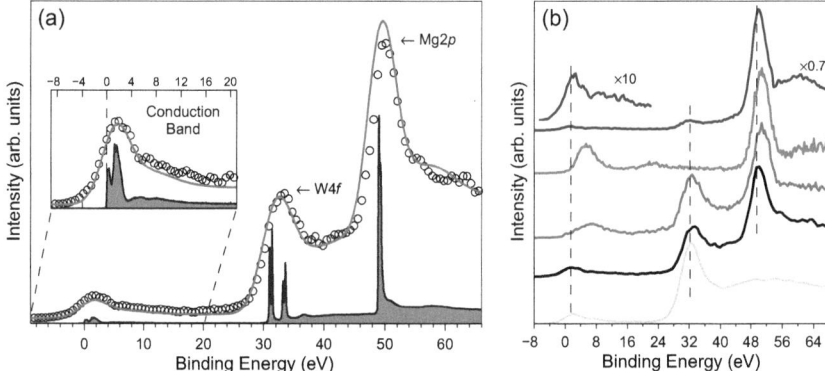

Figure 4.31: Stationary photoemission from Mg adlayers on W(110). (a) Comparison of photoelectron spectra of 1 ML Mg/W(110) obtained with $\hbar\omega = 120$ eV synchrotron radiation (gray shaded area), and with HH cut-off radiation filtered by the 118 eV @ 4.2 eV bandwidth XUV mirror (circles). The solid line corresponds to the synchrotron spectrum convoluted with a 4.4 eV FWHM Gaussian. (b) XUV-only spectra collected from (top to bottom): 4 ML Mg/W(110), magnesium oxide, 1 ML Mg/W(110) with surface contamination, pristine 1 ML Mg/W(110) and clean W(110).

4.3.2 Layer-Resolved Attosecond Streaking

A typical XUV-only spectrum of 1 ML Mg/W(110) excited with HH radiation filtered by the 118 eV @ 4.2 eV bandwidth multilayer mirror is depicted in Fig. 4.31 (a). To ensure a maximum degree of crystalline order, the saturated monolayer is prepared by shortly annealing a W(110) crystal covered with a Mg multilayer to ∼600 K. Despite the large bandwidth of the exciting attosecond pulses, the Mg2p and W4f features remain clearly separated. Compared to the emission strength of the clean surface, the W4f signal is attenuated by more than 40 % but is still peaking sufficiently above the background produced by the inelastically scattered electrons. As expected from the high-resolution photoemission data shown in Fig. 4.30 (c), the line-shape of the conduction band emission exhibits the same asymmetry as observed for pristine W(110). Moreover, the overall spectral shape of the entire photoelectron spectrum is in good agreement with the convoluted synchrotron measurement performed with $\hbar\omega = 120$ eV. The top trace in Fig. 4.31 (b) shows a photoelectron spectrum obtained after evaporation of 4 ML Mg onto the clean W(110) surface, which represents the thickest overlayer investigated in this thesis. All samples with Mg coverages above 1 ML have been annealed to 450 K with 2 K/s, i.e. close to the multilayer desorption temperature, to further improve the crystallinity of the deposited films. With 4 ML Mg on top of the W(110) surface, the W4f emission is severely damped by ∼80 % due to elastic and inelastic collisions suffered by the primary W4f photoelectrons during their propagation through the Mg adlayers. The resultant

Attosecond Dynamics at Metal-Metal-Interfaces

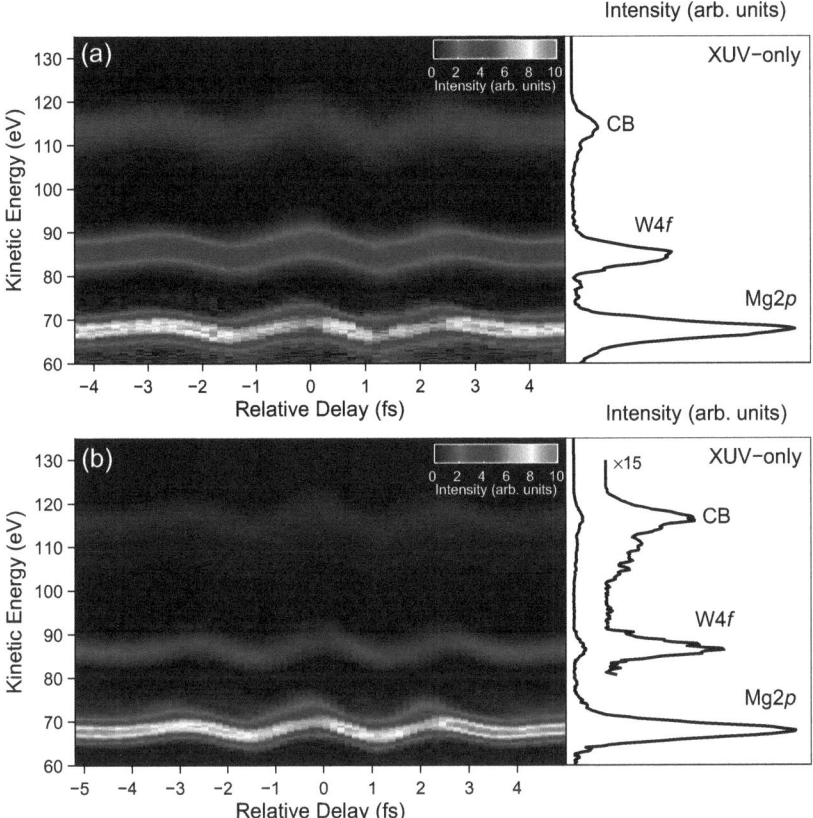

Figure 4.32: Attosecond time-resolved photoemission from Mg/W(110) interfaces. (a) Background-subtracted spectrogram obtained from a Mg monolayer on W(110) with an XUV energy of ∼118 eV and a NIR intensity of $1 \cdot 10^{11}$ W/cm^2. Each of the 56 individual photoelectron spectra composing the spectrogram are averaged over ∼$7.5 \cdot 10^4$ laser shots, resulting in total acquisition time of less than 25 min. (b) Streaking spectrogram collected from W(110) covered by 4 ML magnesium. The individual electron spectra are integrated for ∼$1.2 \cdot 10^5$ laser shots to achieve a sufficient S/N ratio for the strongly attenuated W4f photoemission. No build-up of contamination was detected during the measurement, despite the rather long acquisition time of ∼45 min. Representative background-corrected NIR field-free spectra are depicted on the right hand side of (a) and (b).

photoemission line is comparably weak, and almost submerged in the background signal. For the 4 ML thick overlayer, the photoelectron spectrum already exhibits spectral features characteristic of bulk Mg(0001). Expecially the plasmon losses associated with the primary CB and Mg2p photoemission become visible.

The magnesium monolayer films turned out to be extremely susceptible to contamination effects. In addition to the characteristic changes in the CB line-shape described in Section 4.1.2, the presence of impurities also induced a small shift of the Mg2p photoemission line to ∼1.5 eV higher binding energies which is indicative of a changed oxidation state of the adsorbed Mg atoms (see red trace in Fig. 4.31 (b)) [152]. This becomes even more evident when comparing the corresponding line positions observed in a spectrum collected from epitaxially grown magnesium oxide (magenta trace in Fig. 4.31 (b)). The acquisition time in the streaking experiments had therefore to be further reduced for this system to avoid excessive incorporation of impurity atoms in the Mg films in the course of the measurements. On the other hand, the lower count rate and smaller signal-to-background ratio in the W4f region of the 4 ML spectrum calls for significantly longer integration times to achieve the data quality necessary for a quantitative analysis of the streaking spectrograms.

A background-corrected spectrogram of 1 ML Mg/W(110) is shown in Figure 4.32 (a) together with a representative XUV-only spectrum. The total integration time was limited to ∼25 min which has proven to be a good trade-off between obtaining a sufficient S/N ratio and keeping the concentration of surface contamination below the detection limit during data acquisition. Owing to the high quality of the Mg films, a strong enough NIR intensity could be applied to induce a sufficient modulation of the photoelectron kinetic energies, which enabled a reliable quantitative analysis of the resulting streaking traces. The corresponding spectrogram obtained from 4 ML Mg/W(110) is depicted in Figure 4.32 (b). Typical acquisition times of ∼45 min were necessary to acquire streaking data with an acceptable S/N ratio in the CB and W4f region. Surprisingly, no indication for an oxidation of the Mg overlayer could be detected after completion of the measurement, thereby suggesting a lower surface reactivity for thicker Mg films.

In this context it should be mentioned that a thickness-dependent reactivity of Mg layers towards oxygen was indeed observed recently, and explained by the modulation of the electron density near the Fermi level due to the formation of quantum-well states confined within the Mg adlayers [153, 154]. It is also known that the surface reactivity of thin metal films is strongly influenced by lattice strain. A combination of both effects might therefore be responsible for the different sensitivities of the investigated Mg layers towards surface contamination, especially when considering that the strain induced by the anisotropic lattice mismatch at the W-Mg interface is gradually alleviated with increasing thickness of the Mg film.

Exemplary streaking traces extracted from a spectrogram of 1 ML Mg/W(110) are presented in Fig. 4.33. The center of energies of the electron distribution (shown as crosses) have been evaluated according to Eq. 4.1 within an energy interval of $124 - 100$ eV for

Attosecond Dynamics at Metal-Metal-Interfaces

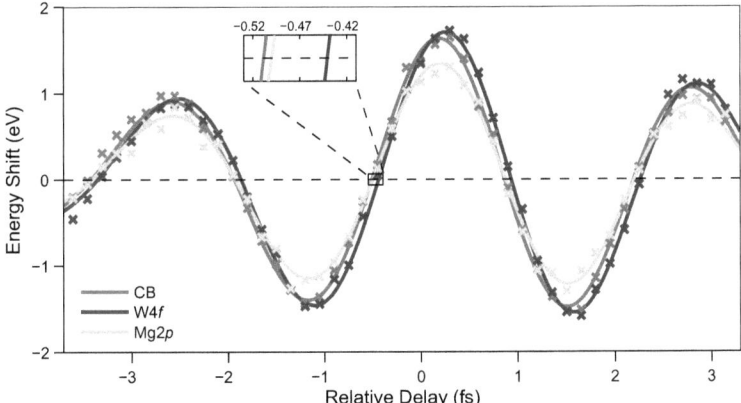

Figure 4.33: COE analysis of the streaked photoelectron distribution from 1 ML Mg on W(110). A global fit to the streaking traces suggests a sizable delay of $\Delta\tau \approx 60-70$ asec between the W4f electrons originating from the W(110) substrate and the Mg2p electrons released from the Mg overlayer, whereas the conduction band and the Mg2p photoelectrons appear to be released almost synchronously into the NIR streaking field.

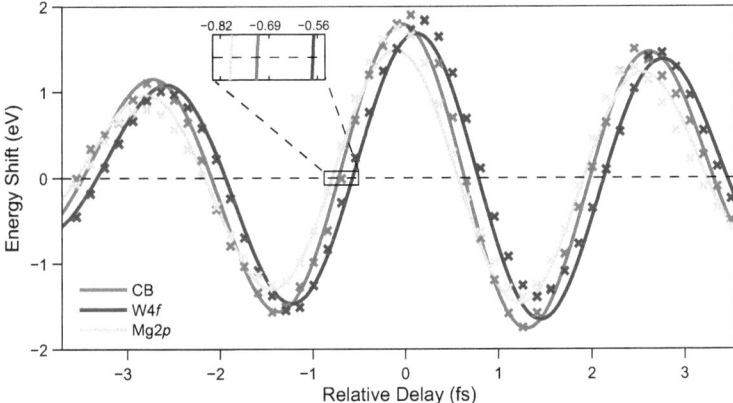

Figure 4.34: The COE analysis of a streaking measurement on 4 ML Mg on W(110) reveals a significantly increased time delay between the W4f and the Mg2p photoemission of the order of $\Delta\tau \approx 200$ asec. For this coverage, also the conduction band electrons appear to be delayed compared to the Mg2p photoelectrons.

Attosecond Dynamics at Metal-Metal-Interfaces

the CB, 98 − 77 eV for the W4f and within the kinetic energy range of 77 − 60 eV for the Mg2p photoelectrons. A straightforward extension of the fitting schemes described in Section 4.1.1 allows the analysis of multiple photoemission lines contained in a streaking spectrogram. The corresponding result of a global fit to the three COE traces is shown as solid lines. A delay of the W4f substrate electrons of the order of $\Delta\tau \approx 60 - 70$ asec with respect to the Mg2p photoelectrons emitted from the monolayer is clearly resolved. The electron emission from the conduction band states, on the other hand, seems to be almost synchronized with the release of the Mg2p electrons into the streaking field.

Figure 4.34 shows the corresponding analysis for 4 ML Mg/W(110). Here, the time shift between the Mg2p and W4f electrons has increased to $\Delta\tau \approx 200$ asec, clearly indicating a pronounced propagation effect experienced by the substrate electrons within the Mg overlayer before being released into the unscreened streaking field. Compared to this, the conduction band electrons are apparently much less affected by this transport phenomenon and exhibit only a small delay of $\Delta\tau \approx 50$ asec.

For a more reliable quantification of these time shifts, all streaking spectrograms were analyzed with a slightly modified version of the TDSE-retrieval. The representation of the electron wave packet describing the W4f and Mg2p photoemission was adopted from the corresponding initial-state parameterization of clean W(110) and bulk Mg(0001) introduced in Section 4.1.1 and 4.1.3, respectively. For the description of the conduction band wave packet released from 1 ML Mg/W(110), the same parameters as for clean W(110) were applied which is justified given the fact that the line-shape of the CB emission is clearly preserved upon adsorption of the Mg monolayer (comp. Fig. 4.30 (c)). This parameterization of the CB electron wave packet was changed only slightly for coverages exceeding 3 ML in order to account for the developing CB plasmon losses and the gradual attenuation of the W5d-dominated contribution to the total CB emission. For this purpose, only the relative intensities of the individual transitions have been adjusted to reproduce corresponding XUV-only spectra. Examples for the TDSE-reconstruction of spectrograms collected from 1 ML and 4 ML Mg/W(110) are depicted in Fig. 4.35 and Fig. 4.36, respectively. The optimized spectrograms describe the measured streaking data sufficiently well. This is further demonstrated in the panels (c)-(e), where measured and reconstructed photoelectron spectra are explicitly compared at selected NIR-XUV pump-probe delays, including the maxima, minima and zero-crossings of the NIR vector potential.

Figure 4.37 and Table 4.2 summarize all results obtained with the TDSE-retrieval for the Mg/W(110) systems as a function of magnesium coverage. Both the time delay of the CB and the W4f electrons are given relative to the Mg2p electron emission from the overlayer. The main objective of this study was to confirm the pronounced contrast in emission times observed for W(110) surfaces covered by 1 ML and 4 ML magnesium, respectively. For the monolayer, an average of 17 independent measurements yields a mean time delay of $\overline{\Delta\tau}_{4f} = 72 \pm 15$ asec for the W4f electrons and of $\overline{\Delta\tau}_{CB} = 10 \pm 19$ asec for the CB electrons, which corroborates the results extracted with the COE method.

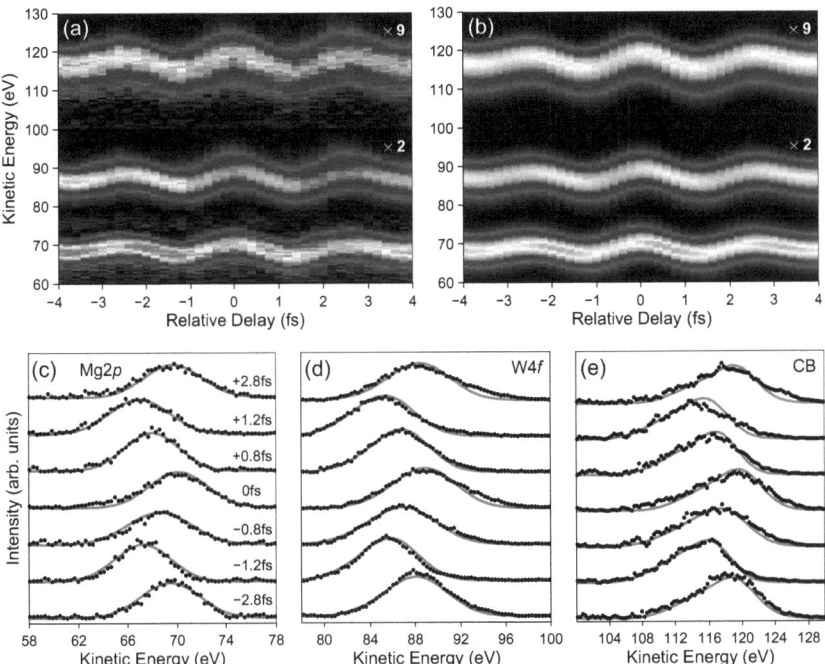

Figure 4.35: TDSE-analysis of attosecond time-resolved photoemission from 1 ML Mg/W(110). (a) Measured streaking spectrogram. (b) Reconstruction of the spectrogram optimized by TDSE-retrieval. The CB and W4f emission is scaled by ×9 and ×2, respectively, to ease comparison. (c)-(e) Comparison of measured (dots) and reconstructed (solod line) photoelectron spectra for selected time delays between the XUV and NIR pulses. The comparison for the Mg2p, W4f and conduction band regions are presented in separate panels and on individual intensity scales.

The corresponding time delays obtained for a coverage of 4 ML magnesium are $\overline{\Delta\tau}_{4f} = 225 \pm 22$ asec and $\overline{\Delta\tau}_{CB} = 29 \pm 15$ asec (average of 5 measurements) which is again in reasonable agreement with the COE result. The tendency of increasing time delay $\Delta\tau_{4f}$ of the W4f electrons with growing thickness of the Mg overlayer clearly persists for all coverages investigated in this thesis, thereby providing unambiguous evidence for a transport-related origin of the observed time shifts. In sharp contrast, the evolution of the conduction band delay $\Delta\tau_{CB}$ as a function of magnesium coverage exhibits obviously a more complex behavior (see Fig. 4.37). Following a strong initial increase, the delay of the conduction band electrons finally stabilizes at ∼50 − 60 asec and seems to even decrease again for coverages above 3 ML. Obviously, this non-monotonic evolution conflicts with a simple explanation in terms of transport effects experienced by the initially localized W5d photoelectrons upon their propagation through the Mg adlayers. A possible explanation

Attosecond Dynamics at Metal-Metal-Interfaces

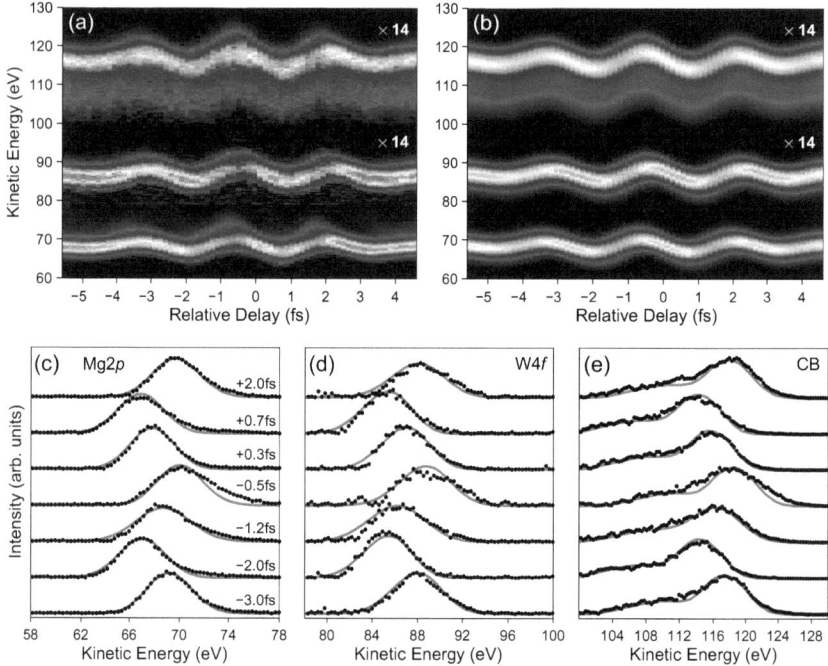

Figure 4.36: Analysis of the streaked attosecond photoemission originating from 4 ML Mg/W(110). (a) Measured spectrogram. (b) Spectrogram reconstructed by the TDSE-retrieval. The CB and W4f emission is scaled by ×14 to simplify comparison. (c)-(e) Comparison of fit (solid line) and data (dots) for selected time delays between the XUV and NIR pulses. The Mg2p, W4f and conduction band emission are displayed in individual panels and on separate intensity scales.

for this unexpected phenomenon will be given in the following subsection.

The relatively large error bars associated with the coverage and shown in Fig. 4.37 are conservative estimates. Although TPD allows in principle for a very precise coverage determination, the use of any spectroscopic techniques ancillary to the streaking measurements had to be limited because of time constraints and the complexity of the main experiment. The Mg coverage was therefore estimated from the deposition time and further cross-checked by comparing the evolution of the Mg2p-W4f intensity ratio in the XUV-only spectra. Nevertheless, this moderate accuracy does not affect the main conclusion which will be drawn in the following subsection from the evolution of the time shifts highlighted in Fig. 4.37. In future experiments, the uncertainty of the Mg coverages can be significantly reduced by performing a TPD analysis for each prepared sample individually.

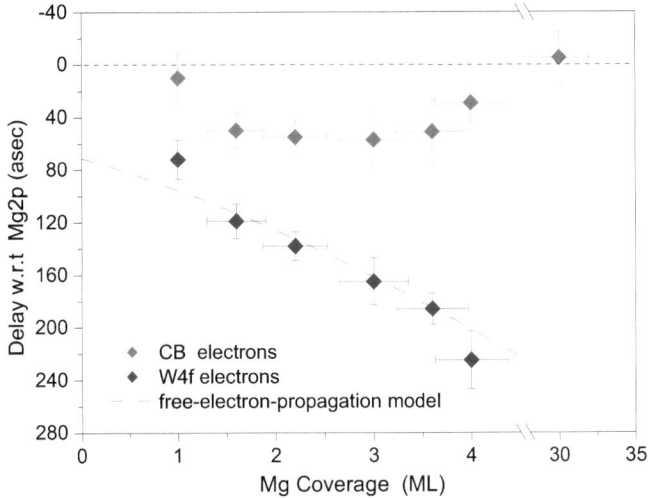

Figure 4.37: Delay of the conduction band (CB) and W4f photoelectrons with respect to the Mg2p photoemission as a function of magnesium coverage. The blue dashed line corresponds to the evolution of the Mg2p-W4f delay expected for free-electron-like propagation of the W4f electrons through the magnesium adlayers (comp. Eq. 4.12).

4.3.3 Discussion

A recurrent problem in the quantitative interpretation of the time shifts observed in streaking experiments on metal surfaces concerns the screening and refraction of the NIR streaking field at the metal-vacuum boundary. Since a realistic theoretical modeling of this effect beyond the Fresnel approximation seems to be unfeasible at the moment, an experimental investigation of this screening phenomenon would be of paramount importance for any future application of the streaking technique to solid-state systems. In this respect, the thickness-dependent time delay between the Mg2p and W4f core-level electrons in the Mg/W(110) systems may serve as an ideal test case. This is because the spatial origin of these photoelectrons is well-defined and disturbing band structure effects may be ignored, at least within the Mg overlayer, due to the free-electron-like final-state dispersion in this simple metal.

Assuming a prompt screening of the streaking field at the magnesium-vacuum interface and free-electron-like propagation of the electrons towards the surface, the difference between the average travel times of electrons released from these two core states $\Delta\tau(d)$ can

101

Attosecond Dynamics at Metal-Metal-Interfaces

Coverage	$\Delta\tau_{CB}$ [asec]	Error($\pm$) [asec]	$\Delta\tau_{4f}$ [asec]	Error($\pm$) [asec]	Scans
1.0ML	10	19	72	15	17
1.6ML	50	13	119	13	2
2.2ML	55	12	138	11	5
3.0ML	57	18	165	18	2
3.6ML	51	24	186	12	5
4.0ML	29	15	225	22	5
30ML	-5	20	—	—	16

Table 4.2: Retrieved time delay of CB ($\Delta\tau_{CB}$) and W4f ($\Delta\tau_{4f}$) electrons relative to the Mg2p photoemission for different coverages of Mg on the W(110) surface. Error margins correspond to standard deviations. For systems with only two data points, the error is approximated by the maximum standard error of the measurements. The data for 30 ML is taken from Section 4.1.3 and is representative of bulk Mg(0001).

be calculated as a function of the Mg overlayer thickness d according to:

$$\Delta\tau(d) = \tau_{R,2p} - \tau_{R,4f} \qquad (4.10)$$

$$= \frac{1}{v_{2p}} \cdot \frac{\int_0^d z\, e^{-z/\lambda_1}\, dz}{\int_0^d e^{-z/\lambda}\, dz} - \frac{1}{v_{4f}} \cdot \left(\frac{\int_d^\infty (z-d)\, e^{-(z-d)/\lambda_2}\, dz}{\int_d^\infty e^{-(z-d)/\lambda_2}\, dz} + d \right) \qquad (4.11)$$

$$= \frac{1}{v_{2p}} \cdot \left(\lambda_1 - \frac{d}{e^{d/\lambda_1} - 1} \right) - \frac{\lambda_2 + d}{v_{4f}}, \qquad (4.12)$$

where $\lambda_1 = 3.7\,\text{Å}$ is the inelastic mean free path of the Mg2p electrons in bulk Mg, $\lambda_2 = 4.3\,\text{Å}$ is the corresponding IMFP of the W4f electrons in W(110) and their free electron velocities v_{2p} and v_{4f} are calculated from the measured electron kinetic energies according to Eq. 2.21. The first term in Eq. 4.11 accounts for the self-attenuation of the Mg2p photoelectron emission within the Mg layer, which limits their average escape depth. The relevant values for λ_1 and λ_2 are inferred from the photon-energy-dependent surface-to-bulk ratio in the Mg2p photoemission (comp. Section 4.1.3). The result of Eq. 4.12 is depicted as blue dashed line in Fig. 4.37 as a function of the overlayer thickness d. The correspondence between the measured time delays and the predictions of this very simple model is astonishingly good. For the necessary conversion of d into a coverage in units of monolayers, a constant interlayer spacing of 2.65 Å was assumed which corresponds to the bulk value of Mg(0001) single crystals.

The fact that this very simple model captures the evolution of the time shift in function of the overlayer thickness is an important experimental finding, since it provides evidence that the kinetic energy modulation of the XUV-induced photoelectrons by the NIR streaking field is indeed negligible before these electrons have escaped into the vacuum. It therefore substantiates the assumption of a vanishing normal component of the NIR electric field within the first atomic layer for metallic samples which represented one of the main uncertainties in the discussion on the relative time delays observed for the clean metal surfaces in Section 4.1. Having confirmed this basic principle of the streaking technique for the Mg/W(110) interfaces, it is worth discussing the time delays measured for 1 ML Mg/W(110) in more detail. Without the Mg overlayer, the CB and W4f photoelectrons arrive with a relative time delay of $\overline{\Delta\tau} = 28 \pm 14$ asec at the surface (see Section 4.1.1). In the presence of the Mg monolayer, the spatial onset of the streaking effect is shifted further away from the Mg-W interface by a distance corresponding to the thickness of the metallic overlayer (~2.5 Å). Under the assumption that the joint CB of 1 ML Mg/W(110) is dominated by the rather localized W5d electrons, one would expect this relative delay to remain unchanged compared to clean W(110). This is because the additional propagation effect experienced inside the Mg adlayer is almost identical for the W5d and W4f photoelectrons, and thus cancels in the calculation of their run-time difference. Apparently, this scenario is not compatible with the measured time delay of ~60 asec between the CB and W4f electron for the Mg monolayer (see Tab. 4.2).

The origin of this discrepancy may be attributed to the different characteristics of the involved electronic states. Whereas the W4f initial-state wave functions can be considered to be strictly spatially confined to the W(110) crystal lattice, the W5d wave functions have the possibility to overlap and interact with valence states of the Mg overlayer to form strongly hybridized electronic states, similar to the situation encountered for the monolayer oxygen on W(110). An independent experimental evidence for this hybridization hypothesis was presented in Section 4.3.1 based on high-resolution synchrotron photoemission spectra. It is highly probable that these newly formed electronic states are substantially delocalized across the Mg-W interface. Consequently, the electron wave packets released from these initial states experience a smaller transport effect within the Mg overlayer than the W4f wave packets emerging from the W(110) substrate which are forced to traverse the full Mg adlayer. In analogy to the O/W(110) system, this translates into an increased time shift between CB and W4f electrons in the streaking experiments. In the same way, such a strong mixing of Mg- and W-related states at the interface naturally explains the small delay of $\overline{\Delta\tau}_{CB} = 10 \pm 19$ asec measured between the Mg2p photoelectrons and the electrons released from these joint CB states in 1 ML Mg/W(110).

On the other hand, the interpretation of the CB time shift for higher magnesium coverages requires a more careful analysis of the composition of the joint Mg-W conduction band states for the different film thicknesses. Since electronic structure calculations are presently not available for these systems, only a qualitative explanation will be given here. The sudden increase of the CB time delay for $1\,\text{ML} \leq \Theta \leq 2\,\text{ML}$ might suggest that the hybridized Mg-W states are spatially confined to the interface region and do not significantly interact with the electronic states in the Mg bilayer. The enhanced time delay can then be interpreted as a propagation effect of electrons originating from the buried interface states, which are still dominating the conduction band photoemission due to the higher excitation cross-section of the W5d-derived states.

For higher coverages, the electronic structure of the Mg films should asymptotically approach the bulk properties of Mg(0001), with the contribution of the W(110) substrate states to the total conduction band emission becoming rapidly attenuated. This transition can be expected to proceed already for rather thin films because of the short inelastic mean free path for electrons in magnesium when excited with an XUV photon energy of $\hbar\omega_x \approx 118$ eV. In Section 4.1.3, it has been demonstrated that CB and Mg2p photoelectrons are emitted almost synchronously from bulk Mg(0001). Therefore, the slight decrease of the delay for $\Theta \geq 3$ ML in Fig. 4.37 may be interpreted as the very beginning of an increasingly Mg-dominated conduction band. This gradual transformation into bulk Mg(0001) is supported by the appearance of plasmon loss lines in the corresponding photoelectron spectra at these coverages. Furthermore, a recent detailed study of the subtle energy shifts occurring in the Mg/W(110) conduction band during epitaxial growth of Mg suggests that the surface state, which is characteristic of bulk Mg(0001), starts to develop only for coverages exceeding $\sim$3 ML [146]. A complete relaxation of the time shift to the Mg(0001) bulk value can therefore not be expected even for a 4 ML thick magnesium film. In this respect it would be interesting to follow the evolution of the relative time delay between the CB and Mg2p states for even higher Mg coverages, which would be feasible with the current experimental setup. Unfortunately, this was not possible within the scope of this thesis because of the limited beamtime allotted to this project.

Figure 4.37 indicates that the time shifts measured for the monolayer differ from the general trend which manifests itself for the thicker magnesium films. Despite the facts that this deviation is rather small for the W4f delay, and a reasonable explanation for the behavior of the conduction band time delay can be given in terms of hybridization of interface states, it is nevertheless important to verify to what extent a possible contamination of the rather sensitive monolayer films may affect these time shifts. The analysis of 9 streaking spectrograms of 1 ML Mg/W(110) exhibiting an extremely high concentration of surface contamination revealed that the presence of impurities in the Mg monolayer tends to even increase the time delay of the W4f core electrons to $\overline{\Delta\tau}_{4f} = 104 \pm 17$ asec, whereas the time delay of the conduction band electrons of $\overline{\Delta\tau}_{CB} = 11 \pm 22$ asec agrees within the error of the measurement with the corresponding mean value derived from clean interfaces. The small deviation of the W4f time delay from the free-electron-like propagation model in Fig 4.37 is therefore *not* indicative of a residual contamination effect in the monolayer samples. A detailed quantitative interpretation of this contamination-induced effect will not be attempted here. It is likely to be rather complex since the oxidation of the Mg films by oxygen-containing impurities will not only significantly change the morphology of the adlayer, but also its electronic and optical properties [155].

4.4 Attosecond Streaking in Dielectrics: Xe/W(110)

The streaking experiments presented so far have exclusively been performed on metallic samples. In these experiments, the component of the NIR field normal to the surface can be assumed to be substantially weaker in the interior of the solid than in vacuum, which is a general consequence of the optical properties of metals. This assumption was corroborated in the previous section, where increasing time shifts with growing thickness of a metallic overlayer were found to agree nicely with a simple transport model for the excited photoelectrons. The general foundation of this transport phenomenon is the spatial separation between the generation of these photoelectrons inside the metal, and their subsequent interaction with the NIR streaking field in vacuum.

Conversely, these propagation effects should be less pronounced in dielectrics with a refractive index closer to unity. In these materials, the screening of the NIR field is less efficient, and the kinetic energies of photoelectrons excited within this material will be almost instantaneously accelerated by the streaking field, similar to experiments with isolated atoms in the gas phase. An interesting application of this phenomenon would be the measurement of average photoelectron travel times τ_R towards the surface of a metal. Obviously, such absolute propagation times are more informative than the relative time delays $\Delta\tau$, which always entail the risk of losing information due to balancing effects between the propagation of the two types of photoelectrons involved in the measurement. Therefore, theoretical models for attosecond photoemission may be more efficiently developed and verified based on experimental values of τ_R [22, 61, 63, 124, 144].

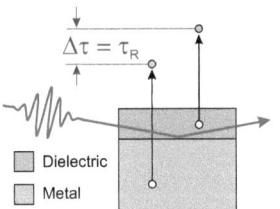

Figure 4.38:
Experimental approach for the investigation of absolute electron travel times τ_R in metals. Photoelectrons released from a dielectric monolayer are assumed to respond immediately to the NIR streaking field, and therefore provide a reference for the measurement of the run-times τ_R for substrate electrons to the metal-dielectric interface. Under these assumptions, the relative time delay $\Delta\tau$ between substrate and overlayer photoelectrons equals τ_R.

First experimental attempts to gain access to these quantities will be discussed in this section. The ideal experimental scenario is sketched in Fig. 4.38. Electrons emitted from a dielectric overlayer serve as a reference for the absolute timing of the primary photoelectrons released from a metal substrate. Because of its low refractive index, the NIR streaking field already acts on the electrons when they arrive at the metal-dielectric interface. In contrast to the metal films investigated in the previous section, this should minimize residual transport effects within the overlayer, and reveal the absolute travel time τ_R for photoelectrons propagating in the metal substrate.

In consideration of the usual constraints concerning suitable materials for streaking experiments, adsorbed xenon (Xe) on W(110) was identified as a test system to explore the

Attosecond Streaking in Dielectrics: Xe/W(110)

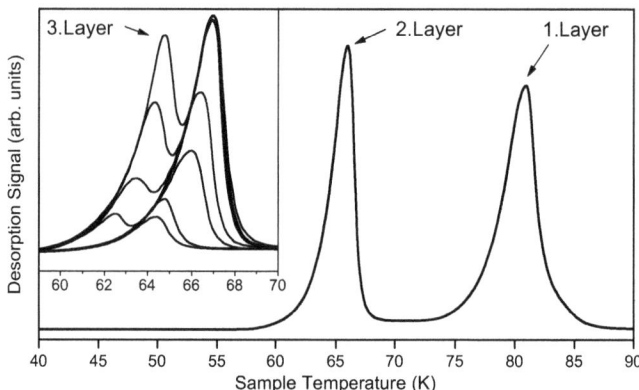

Figure 4.39: Thermal desorption spectrum of 2 ML Xenon on W(110) recorded with a temperature rate of 1 K/s. The inset shows the evolution of the low-temperature desorption signal for initial coverages 1 ML $< \Theta \leq 3$ ML.

possibility of measuring absolute propagation times in attosecond photoemission from metals. The refractive index of solidified xenon is only $\tilde{n} \approx 1.5$ in the NIR spectral range and exhibits negligible absorption due to the large band gap of ∼9 eV eV and the high energy necessary for excitonic excitations (∼7 eV for the surface exciton) [156, 157, 158]. In addition, because of its closed-shell electron configuration, the rare gas xenon atoms can be expected to interact only weakly with the electronic system of the W(110) substrate. Thus, in contrast to adsorbates like oxygen and magnesium studied in the previous sections, complications arising due to the hybridization of electronic states at the interface can be avoided.

4.4.1 Sample Preparation and Characterization

For the deposition of xenon, the clean W(110) crystal was cooled to ∼30 K and exposed to high-purity xenon gas admitted to the surface through a micro-capillary doser. Figure 4.39 shows the thermal desorption signal obtained from the W(110) surface with a temperature rate of 1 K/s after exposure to a Xe dosage corresponding to a final coverage of $\Theta = 2$ ML. The TD trace features two well-resolved maxima related to the desorption of the monolayer at ∼80 K and the Xe bilayer at ∼65 K. Xenon desorption for initial coverages in the range of 1 ML $< \Theta \leq 3$ ML is shown in the inset of Fig. 4.39. For increasing Θ, the bilayer desorption signal eventually saturates and a new desorption peak evolves at ∼3 K lower temperature, which can be ascribed to Xe atoms desorbing from the third atomic layer. The sequential appearance of these desorption states indicates a layer-by-layer

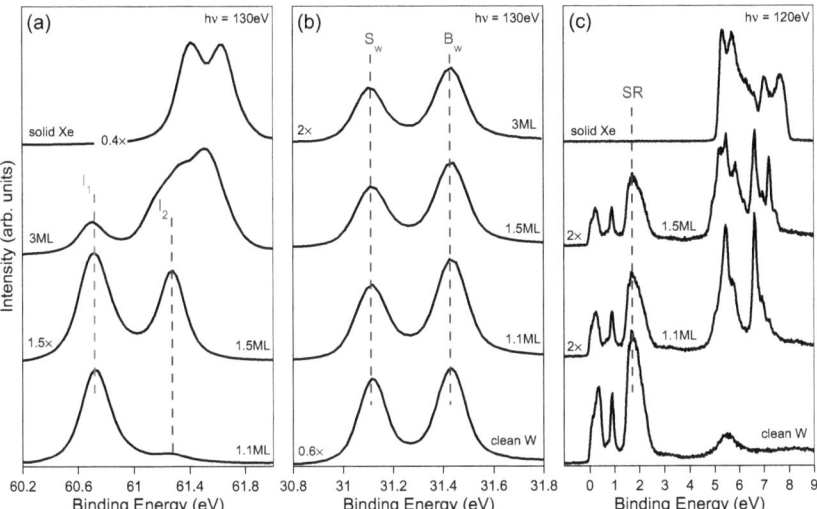

Figure 4.40: Adsorption of Xe on W(110) studied with high-resolution synchrotron photoelectron spectroscopy (acceptance angle ≈ ±0.5°). Changes in (a) the Xe$4d_{5/2}$, (b) the W$4f_{7/2}$ and (c) the conduction band emission are followed as function of Xe coverage. Corresponding spectra of clean W(110) and solid Xenon are shown for comparison.

growth of Xe on W(110) in accordance with an early study by Opila et al. [159].

The weak interaction of the Xe adatoms with the W(110) substrate is not only reflected in the low monolayer desorption temperature, but also in the high-resolution synchrotron photoemission spectra summarized in Fig. 4.40. Panel (b) depicts the evolution of the W$4f_{7/2}$ emission for Xe coverages up to ∼3 ML. Except for the gradual attenuation of the overall intensity with increasing thickness of the Xe overlayer, the line-shapes of both the surface (S_W) and the bulk (B_W) component are strictly preserved. This corroborates the assumption that the charge distribution at the W(110) surface is not significantly affected upon the adsorption of the Xe atoms. The conduction band spectra presented in panel (c) further substantiate this conclusion. Especially, the sharp spectral features in the binding energy range of 0−3 eV below E_F, which is dominated by the W(110) surface resonance (SR), remain essentially unperturbed. The spin-orbit split Xe$5p$ valence levels appear at ∼5.5 eV and ∼6.5 eV binding energy, respectively. These states start dominating the conduction band emission already in the monolayer coverage regime.

The strongest evidence for the high quality of the adsorbed xenon films is provided by the layer-dependent binding energy shifts observed for the Xe$4d_{5/2}$ core levels in panel (a). The monolayer spectrum is characterized by a single emission line I_1. Upon further Xe deposition, this component is rapidly attenuated and a new emission line I_2, which can be assigned to photoemission from Xe atoms adsorbed on top of the saturated monolayer, develops at ∼550 meV higher binding energy. For $\Theta > 2$ ML, additional components related to the photoemission from the third and forth Xe layer begin to develop on the high binding energy side of I_2. All these energy shifts originate mainly from the distance-dependent final-state polarization screening of the Xe$4d^{-1}$ core holes mediated by the image charge potential induced by the electrons in the metal substrate [160]. Similar energy shifts can also be observed for the Xe$5p$ valence photoemission in panel (c). However, these states cannot be considered as atomic-like energy levels, since the lateral interactions between neighboring xenon atoms give rise to the formation of two-dimensional energy bands within the xenon layers [161]. Further examination of the I_1 line position reveals a gradual shift to lower binding energies of up to ∼50 meV for higher Xe coverages. This can be explained by the additional dielectric screening of the monolayer Xe$4d^{-1}$ core holes provided by the surrounding xenon matrix [162].

The top trace in Fig. 4.40 (a) was obtained from a W(110) surface covered with ∼10 ML xenon. At this coverage, due to the high surface sensitivity, the spectrum is already representative of bulk xenon. On the other hand, the insulating xenon overlayer is still thin enough to avoid charging effects, which allows the acquisition of high-resolution photoelectron spectra. As a consequence, the ∼240 meV splitting of the Xe$4d$ emission into bulk and surface contributions is clearly resolved. This energy difference between the bulk and surface components is in very good agreement with the value of 250 meV reported for epitaxially grown Xe(111) crystals by Kaindl *et al.* [163].

The experimental results presented above are prototypical for the adsorption of xenon on closed-packed metal surfaces and provide strong evidence for the epitaxial growth of Xe(111) on W(110) [164]. The Xe/W(110) system therefore meets all requirements necessary for well-defined attosecond streaking experiments.

4.4.2 Xenon Monolayer

A typical XUV-only spectrum of 1 ML Xe/W(110) is depicted in Figure 4.41 (a). The layer was prepared by adsorbing Xe in excess of 1 ML and subsequently annealing the W(110) crystal to 70 K with 1 K/s. In this way, a densely-packed centered-rectangular Xe superlattice is formed on the W(110) surface [165]. For the excitation energy $\hbar\omega_x \approx$ 118 eV, the Xe$4d$ photoemission line is sufficiently separated from the $N_{4,3}O_{2,3}O_{2,3}$ Auger electrons, despite the 4.2 eV bandwidth of the filtered XUV radiation (see Fig. 4.41 (b)). These Auger electrons result from the Xe$4d^{-1}$ core-hole decay involving two electrons in the Xe$5p$ valence levels, and appear at constant kinetic energies in the range of 30−40 eV.

Due to their small energy separation, the W$5d$ and Xe$5p$ states merge into a single asymmetric CB peak extending to energies of up to ∼12 eV below E_F for the broadband attosecond excitation, which complicates the analysis of this spectral feature (see also Appendix C). On the other hand, the photoemission from the W$4f$ and Xe$4d$ levels is

Attosecond Streaking in Dielectrics: Xe/W(110)

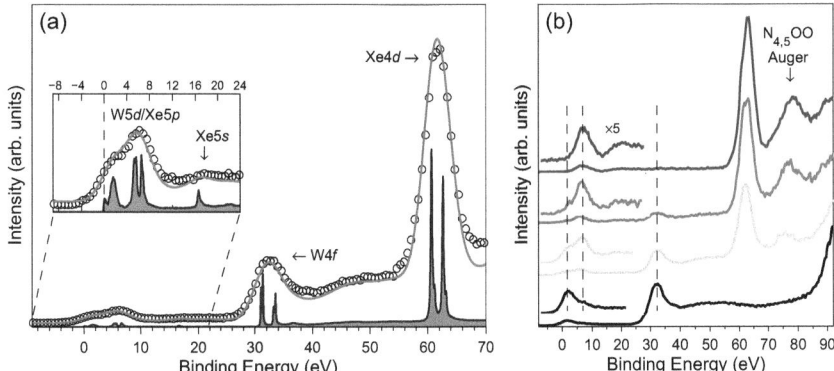

Figure 4.41: Stationary photoemission from 1 ML Xe/W(110). (a) A photoelectron spectrum acquired with $\hbar\omega = 120$ eV synchrotron radiation (gray shaded area) is compared to a corresponding NIR field-free spectrum obtained with sub-fs XUV pulses filtered by the 118 eV @ 4.2 eV multilayer mirror from the HH cut-off continuum. The solid line represents a convolution of the synchrotron spectrum with a 4.4 eV FWHM Gaussian. (b) Comparison of XUV-only spectra collected from (top to bottom): solid xenon, 3.5 ML Xe/W(110), 1 ML Xe/W(110) and clean W(110).

not overlapping with any other emission line. This opens up the possibility to retrieve the average propagation time of the W4f electrons towards the W(110) surface by measuring their relative time delay with respect to the Xe4d photoelectrons in an attosecond streaking experiment according to the scenario sketched in Fig. 4.38.

The combination of the large photo-ionization cross-section of the Xe4d subshell, and the moderate attenuation (-40 %) of the W4f emission upon deposition of 1 ML xenon enabled the acquisition of high-quality streaking spectrograms, as presented in Figure 4.42, in less than 40 min. During this time no decrease in the Xe4d signal strength could be observed for NIR intensities $< 10^{11}$ W/cm^2, which would be indicative of laser-induced desorption of the xenon adatoms from the W(110) surface. On the other hand, these low NIR intensities translate into a relatively weak modulation of the photoelectron kinetic energies, which renders the extraction of timing information from the streaking spectrograms more challenging. Attempts to apply larger streaking field strengths and balancing the NIR-induced depletion of the Xe monolayer by continuously dosing Xe gas onto the W(110) surface through the background pressure were not successful. The difficulties in adjusting the partial xenon pressure to exactly compensate the laser-induced desorption rate resulted in intolerable variations of the Xe coverage during the streaking measurements, even when the sample temperature was stabilized at 70 K, i.e. between the monolayer and bilayer desorption temperature.

Despite the low modulation depth, a quantitative analysis of the streaking spectrograms was nevertheless possible owing to the excellent S/N ratio of the photoelectron spectra.

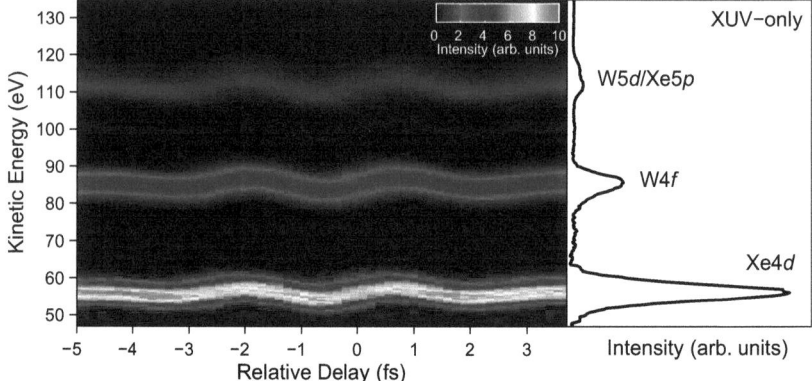

Figure 4.42: Background-corrected streaking spectrogram obtained from 1 ML Xe/W(110) with an XUV energy of $\hbar\omega \approx 118$ eV and a NIR intensity of $7 \cdot 10^{10}$ W/cm². The relative NIR-XUV delay was varied in steps of 150 asec. Photoelectron spectra were accumulated over $\sim 1.1 \cdot 10^5$ laser shots at each delay step.

A representative COE analysis is shown in Fig. 4.43. The first moments of the CB and W4f emission were calculated in rather narrow energy intervals of 120 − 105 eV and 95 − 75 eV to minimize the influence of the Xe5s photoemission which is centered at a kinetic energy of ∼100 eV (better visible in Fig. 4.44 (a)). The corresponding COEs of the Xe4d photoemission were evaluated in the kinetic energy range of 65−45 eV. A global fit to the resultant streaking traces reveals a time delay of $\Delta\tau \approx 100$ asec between the emission of the Xe4d and the W4f photoelectrons.

The analysis of the streaking spectrogram with the TDSE-retrieval is summarized in Fig. 4.44. Here, the spin-orbit splitting of 2 eV and the intensity ratio $I_{4d_{5/2}}/I_{4d_{3/2}} = 1.3$ derived from synchrotron photoemission spectra[9] were considered in the description of the Xe4d electron wave packet. In accordance with the synchrotron photoemission spectrum in Fig. 4.40 (c), the contribution of the Xe5p states to the CB wave packet was accounted for by expanding the parameterization of the CB wave packet of clean W(110) with two additional initial states at 5.5 eV and 6.5 eV binding energy, respectively. The absolute intensities of these two transitions were adjusted to reproduce the measured XUV-only spectrum of 1 ML Xe/W(110). The final simulated spectrogram optimized by the TDSE-routine under the constraints of this initial-state configuration is in good agreement with the experimental streaking data (see Fig. 4.44 (b)).

By comparing Fig. 4.44 (c)-(e) it becomes evident that the fits to the laser-dressed Xe4d and W4f core-level emission are of slightly better quality than for the CB emission. It

---

[9]The same deviation from the statistical value of 1.5 was reported for Xe in the gas phase [166].

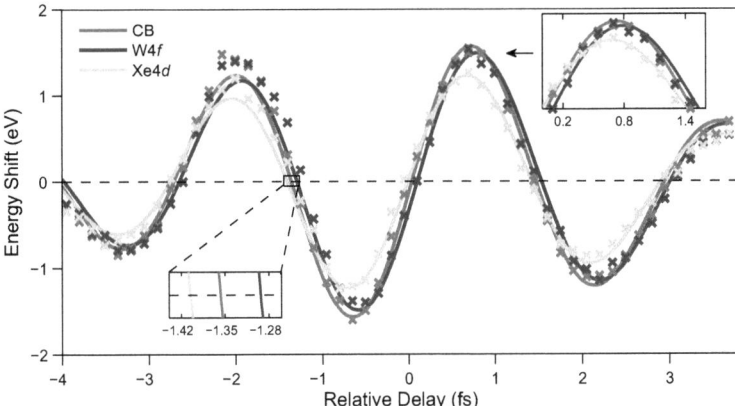

Figure 4.43: The COE analysis of a streaking measurement of 1 ML Xe/W(110) suggests a delay between the Xe4d and W4f core-level photoelectrons of $\Delta\tau \approx 100$ asec.

should be noted that the reconstruction of the streaked CB electron distribution can be improved by allowing a temporal delay between the electrons released from the W5d- and Xe5d-dominated energy levels. In contrast to the mixed O2p/W5d conduction band discussed in Section 4.2, such a delay can be expected for the 1 ML Xe/W(110) system since the spatial localization of the Xe5p and W4f wave functions is preserved due to the absence of hybridization at the Xe/W(110) interface. However, the implementation of this additional optimization parameter resulted in rather unstable fits. It was therefore not possible to reach a definite conclusion concerning the existence of such a time delay. Quite generally, such a time shift between two spectrally not fully resolved electron distributions will give rise to a line-shape modulation in the spectrograms that is similar to the modulations expected for a chirped wave packet. Tests with simulated streaking spectrograms suggest that time shifts down to ∼10 asec between the Xe5p and W4f photoemission might be reliably extracted in future experiments, provided that nearly transform-limited attosecond pulses are employed (see Appendix C).

In this respect, the analysis of the streaked core-level electron wave packets is unambiguous. The evaluation of 7 streaking measurements with the TSDE-retrieval yields a mean time delay of $\overline{\Delta\tau} = 95 \pm 10$ asec between the photoemission from the W4f and Xe4d core states, in excellent agreement with the COE result. Surprisingly, this time shift is even larger than the Mg2p-W4f delay of ∼72 asec extracted from the streaking spectrograms of 1 ML Mg/W(110) in Section 4.3, which might be considered as an upper limit for the average travel times of the W4f electrons in W(110). Further, even within the free-electron-like propagation model a smaller travel time of $\tau_R = 70-80$ asec is expected for the W4f electrons (see Tab. 4.1 on page 71). These discrepancies therefore render an

Attosecond Streaking in Dielectrics: Xe/W(110)

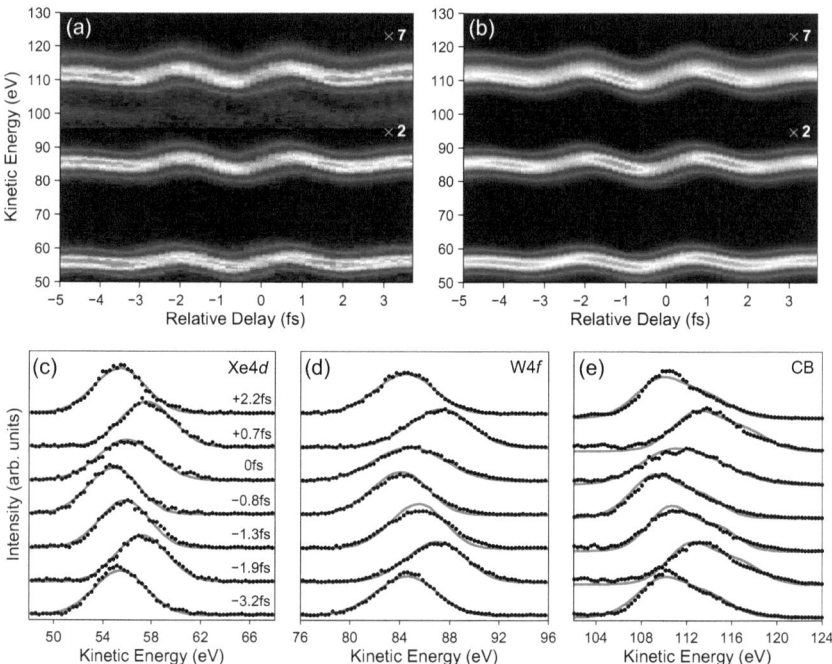

Figure 4.44: Example for the TDSE-analysis of the streaked photoemission collected from 1 ML Xe/W(110). (a) Measured streaking spectrogram. (b) Fit result of the TDSE-retrieval. The CB and W4f regions are scaled to match the intensity of the Xe4d emission. (c)-(e) Comparison of measured and reconstructed photoelectron spectra for selected XUV-NIR delays. The weak Xe5s emission at ∼100 eV kinetic energy was not included in the fitting procedure.

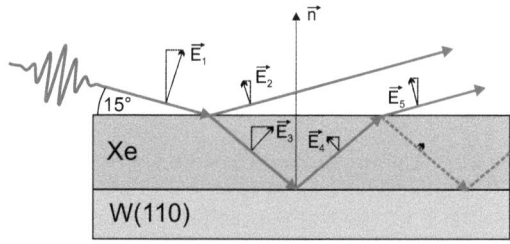

Figure 4.45: Estimate of the effective streaking field strengths at the xenon-W(110) interface based on Fresnel's equations. The total electric field in vacuum is given by $E_{Vac} \approx E_1 + E_2 + E_5$, whereas the electric field inside the xenon layer can be approximated as $E_{Xe} \approx E_3 + E_4$. The application of Fresnel equations with the complex indices of refraction $\tilde{n}_{Xe} \approx 1.5$ and $\tilde{n}_W = 3.8 + 2.7i$ ($\lambda_L = 750$ nm) yields a ratio of $E_{Xe,\perp}/E_{Vac,\perp} \approx 0.3$ for the corresponding field components normal to the surface.

Attosecond Streaking in Dielectrics: Xe/W(110)

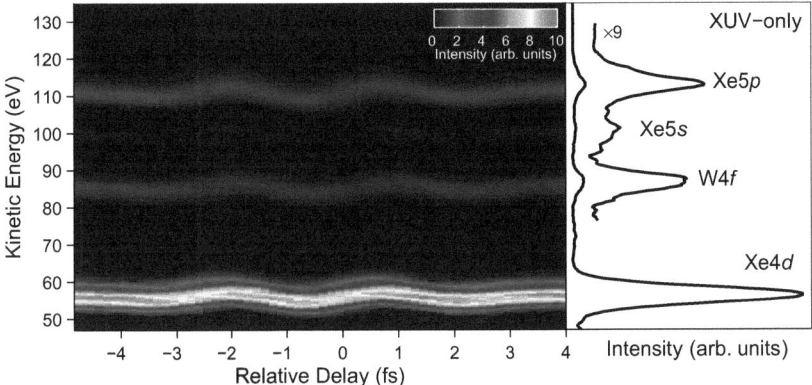

Figure 4.46: Background-corrected streaking spectrogram of 3.5 ML Xe/W(110) acquired under identical experimental condition as the monolayer measurement presented in Fig. 4.42. A photoelectron spectrum obtained in the absence of the NIR streaking field is depicted on the right hand side.

interpretation of the measured W4f-Xe4d delay as the average run-time τ_R of the W4f photoelectrons towards the W(110) surface questionable.

The larger time shift for the Xe monolayer may at least partly be attributed to residual propagation effects of the W4f electrons inside the xenon overlayer. Especially when considering the larger thickness of a Xe monolayer of ∼2.9 Å, which can be calculated from the relative attenuation of the W4f emission and the measured electron inelastic mean free path in condensed Xe (comp. Fig. 4.51 on page 121). When the xenon overlayer is modeled as a dielectric continuum with a refractive index $\tilde{n} \approx 1.5$ [156, 157], the NIR field component normal to the surface is calculated to be ∼70% smaller inside the Xe layer compared to the corresponding field component in vacuum. This value is derived by applying the Fresnel equations, taking into account both the reflected and transmitted electric fields at the vacuum-xenon and xenon-tungsten interface (see Fig. 4.45). It is therefore quite possible that this remaining streaking field inside the xenon overlayer might be too weak to fully eliminate temporal effects related to the propagation of the W4f electron through the xenon adlayer.

4.4.3 Coverage Dependency

The existence of residual transport effects in the adsorbed xenon layers can be verified experimentally by measuring streaking time shifts as a function of the Xe overlayer thickness. A NIR field-free spectrum of 3.5 ML Xe/W(110) is depicted as red line in Fig. 4.41. This sample was prepared by exposing the clean W(110) surface to a roughly predetermined

Attosecond Streaking in Dielectrics: Xe/W(110)

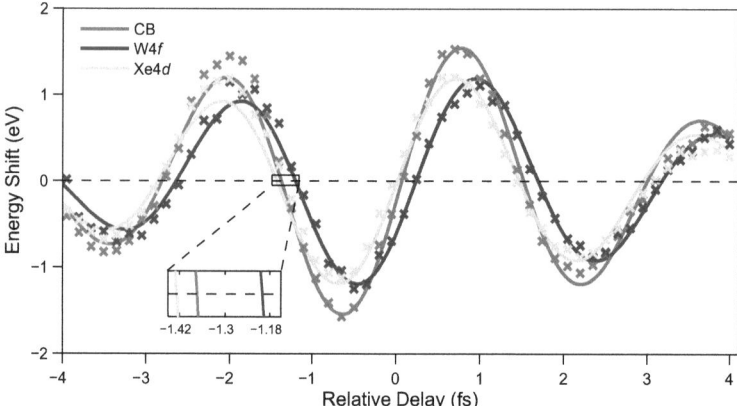

Figure 4.47: COE analysis of the 3.5 ML Xe/W(110) streaking spectrogram depicted in Fig. 4.46. The increased Xe4d-W4f time shift of $\Delta \tau \approx 200$ asec is indicative of a pronounced propagation effect experienced by the substrate W4f electrons traversing the Xenon overlayer.

dosage of xenon gas with the sample temperature stabilized at 30 K. The W(110) crystal was further annealed to 55 K to allow for the best ordering of the atoms in the xenon film. The exact coverage was established in a subsequent thermal desorption experiment.

For this coverage of 3.5 ML xenon, the CB region is almost completely dominated by the Xe5p valence states due to the diminishing contribution of the W5d states which are efficiently scattered upon propagation through the xenon adlayers. The spectral shape of the resultant CB feature therefore closely resembles the valence band of solid xenon (blue curve in Fig. 4.41). In addition, the W4f emission strength is significantly reduced by -80 % compared to the photoemission signal collected from the clean W(110) surface.

Figure 4.46 shows the result of a streaking experiment performed on this system. The NIR intensity and integration time was set to similar values as applied in the monolayer experiments. Evidently, the Xe4d/W4f intensity ratio remains constant throughout the entire streaking measurement, which proves the absence of laser-induced desorption even for these less tightly bound Xe adlayers. The COE analysis of the extracted streaking traces is presented in Fig. 4.47. A significantly increased time delay of $\Delta \tau \approx 200$ asec for the W4f substrate photoelectrons with respect to the Xe4d photoemission is clearly discernible. As will be shown below, the magnitude of this time shift provides strong evidence for the prevalence of transport effects in time-resolved attosecond photoemission from condensed xenon.

The TDSE-analysis of the streaking spectrogram is summarized in Fig. 4.48. In order to account for the dominating character of the Xe5p states in the CB emission, the am-

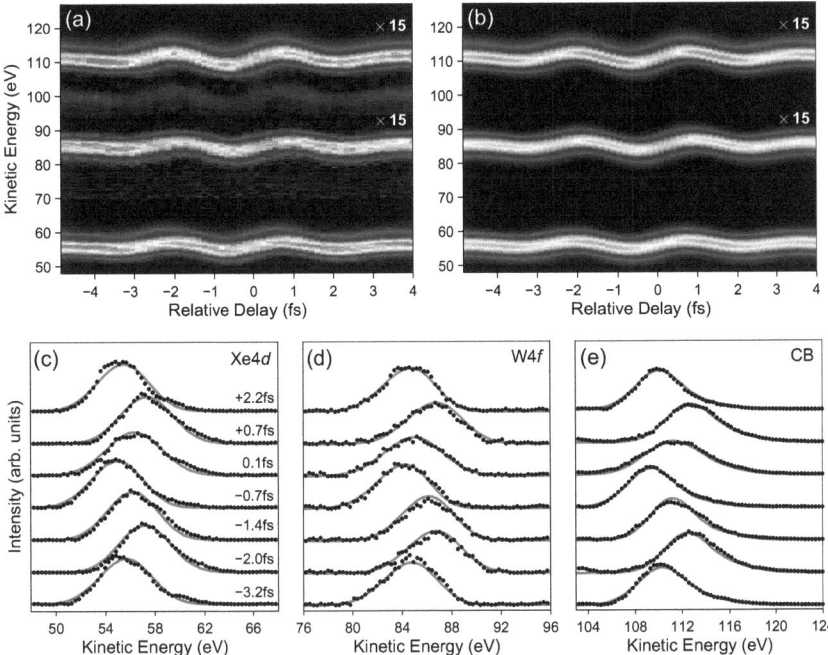

Figure 4.48: TDSE-analysis of the streaked attosecond photoemission originating from 3.5 ML Xe/W(110). (a) Measured spectrogram. (b) Spectrogram reconstructed by the TDSE-retrieval. To simplify comparison, the signal in the kinetic energy range of 70−125 eV is scaled by ×15 in both panels. (c)-(e) Comparison of reconstructed (solid line) and measured (dots) photoelectron spectra for different relative delays between the XUV-pump and the NIR-probe pulses.

plitudes of the individual wave packets representing these states were increased at the expense of the remaining W5p-related states to match the experimental NIR field-free spectrum. In analogy to the evaluation of the monolayer spectrograms, the photoemission from the Xe5s states is ignored in the reconstruction procedure. The excellent quality of the fitted spectrogram is demonstrated in Fig. 4.48 (c)-(e). Both the centers and line-shapes of all three photoemission peaks are well reproduced over the entire range of XUV-NIR delays. Furthermore, the Xe4d-W4f delay of $\Delta\tau = 188 \pm 9$ asec retrieved from this reconstruction is in good agreement with the time shift extracted from the same spectrogram with the COE method. Additional streaking measurements performed on W(110) covered with a Xe bilayer indicate a smaller delay of $\Delta\tau = 150 \pm 8$ asec, which strongly points towards a thickness-dependent transport effect experienced by photoelectrons propagating through the xenon layers (see Tab. 4.4).

Coverage	d [Å]	$\Delta\tau_{30\%}$ [asec]	$\Delta\tau_{0\%}$ [asec]	$\Delta\tau_m$ [asec]	Error($\pm$)
1.0ML	2.9	85	92	95	10
2.0ML	6.4	111	122	150	8
3.5ML	11.6	160	183	188	9

Table 4.3: Time delay between Xe4d and W4f photoelectrons for xenon layers of different thickness d on the W(110) substrate. $\Delta\tau_m$: measured delay; $\Delta\tau_{0\%}$: delay predicted by the free-electron-like propagation model with vanishing streaking field in xenon; $\Delta\tau_{30\%}$: delay predicted by the free-electron-like propagation model with a residual streaking field strength of 30 % inside the xenon film (compared to the vaccum field strength).

A quantitative modeling of the coverage dependency of these time delays, even in terms of free-electron-like propagation, is slightly more complicated than for the magnesium layers discussed in Section 4.3. This is mainly because of the spatially inhomogeneous electric field seen by the W4f photoelectrons on their way from the substrate, through the xenon layers and during their escape into vacuum where they finally experience the full strength of the streaking field. Furthermore, also the Xe4d electrons will experience a spatially varying streaking field when they cross the xenon-vacuum interface. Assuming zero streaking field strength in *both* the W(110) substrate and the xenon overlayer, the delay between these two core-level electrons can be calculated using the same free-electron transport model which was invoked in Section 4.3.3 to estimate the time shifts for the magnesium-covered tungsten crystal (Eq. 4.12 on page 102). The inelastic mean free paths of electrons in xenon, which are needed as input parameters for the propagation model, were derived experimentally from synchrotron photoemission data (comp. Fig. 4.51). In the calculation of the effective layer thicknesses, the Xe-Xe interlayer spacings for $\Theta > 1$ are assumed to be identical to the corresponding values reported for Xe(111) single crystals [163]. The resultant time delays $\Delta\tau_{0\%}$ are compared to the experiment in Tab. 4.3. Obviously, this strongly simplifying model already predicts time shifts which are in reasonable agreement with the experimental results.

For a more realistic description, the influence of remaining 30 % of normal streaking field component inside the xenon layer has to be investigated. In order to get an estimate for the streaking time delay developing between photoelectrons that are generated and traveling through such a system, both the temporal and spatial dependencies of the effective streaking field component normal to the surface $E_L(z,t)$ have to be explicitly accounted for (z denotes the spatial coordinate along the surface normal). For this purpose, streaking traces $S(\tau)$ were simulated by numerically calculating for different relative NIR-XUV delays τ the electron trajectories $z(t,\tau)$ within the Verlet approximation [167]:

$$z(t+\delta t,\tau) + z(t-\delta t,\tau) = 2z(t,\tau) + \frac{e\,E_L(z,t-\tau)}{m_e}\delta t^2 + \mathcal{O}(\delta t^4) \quad (t \geq \tau) \quad (4.13)$$

The time step discretization δt for the numerical calculations was set to 1 asec. The energy shift $S(\tau)$ induced by the streaking field for a specific NIR-XUV time delay τ is given by the additional kinetic energy accumulated by the electrons after the NIR pulse is over:

$$S(\tau) = \frac{1}{2} m_e \left(\frac{z(t,\tau) - z(t-\delta t, \tau)}{\delta t} - v_{free} \right)^2. \tag{4.14}$$

In analogy to Eq. 4.12, the initial starting positions $z(t=0)$ are determined by the mean escape depth of the W4f electrons in W(110) and of the Xe4d electrons in solid xenon. The electrons start propagating with their free electron velocities v_{free} and are accelerated in a piecewise spatially constant electric field, whose field strength is set to zero inside the W(110) crystal, to 30% inside the xenon layers and to 100% in vacuum, in accordance with the extimate based on Fresnel equations (comp. Fig 4.45). As usual, the velocities v_{free} have been caculated according to Eq. 2.21. For solid xenon, the bottom of the conduction band is only 0.4 eV below the vaccum level [158]. A correction of v_{free} due to the inner potential V_0 can therefore be neglected for electrons propagating in the xenon layers[10].

The corresponding time delays $\Delta\tau_{30\%}$ extracted from these simulated streaking traces are also listed in Tab. 4.3 for the different xenon layer thicknesses. As expected, these time shifts are systematically smaller than the time delays $\Delta\tau_{0\%}$ derived for a vanishing streaking field in the xenon layer, and their agreement with the experiment is therefore slightly worse. Nevertheless, within the uncertainty of the measurements, both transport models reproduce the observed experimental trend sufficiently well, thus providing strong evidence for the prevalence of electron propagation-induced temporal effects in the attosecond photoemission from xenon, due to the reduced normal component of the NIR streaking field strength inside the adsorbed films. It should be noted that within the applied Fresnel approximation, this reduction of the relevant field component arises mainly from the refraction of the NIR light at the interface. Detailed future calculations of this electric field, taking the microscopic structure of the adsorbate layer into account, will have to prove (or disprove) the validity of the simple interpretation presented here.

4.4.4 Streaking of Solid and Gas-Phase Xenon

The streaking experiments on xenon-covered W(110) demonstrate that experimental access to absolute electron travel times is hampered by propagation effects occurring even within dielectric adlayers. In principle, the influence of these transport effects can be minimized in future streaking measurements when sub-monolayer coverages of xenon are employed. Tracking the resulting time shifts as a function of coverage may allow to determine the "time delay" for vanishing adlayer coverage by extrapolation. However, a more fundamental problem that arises in the context of absolute timing measurements concerns the possibility of a time lag between the absorption of the XUV pulse and the ejection of the photoelectron from a single atom.

---

[10]Changes in the work function upon xenon adsorption are < 1 eV and therefore also neglegible [168].

Attosecond Streaking in Dielectrics: Xe/W(110)

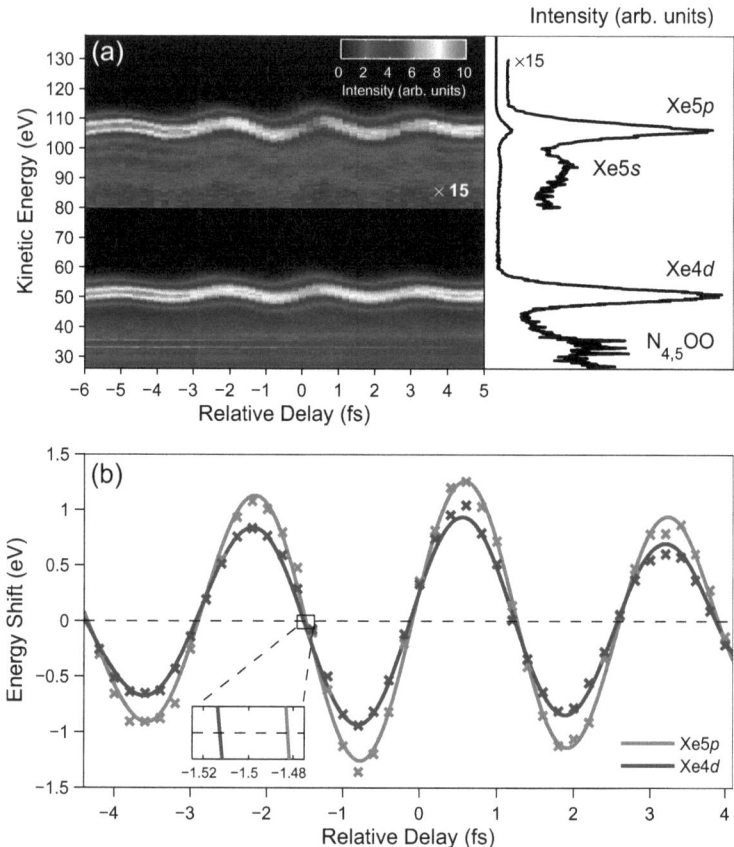

Figure 4.49: Attosecond streaking of gas-phase xenon. (a) Streaking spectrogram obtained with sub-fs XUV pulses centered at 118 eV and a NIR dressing field intensity of $1 \cdot 10^{11}$ W/cm². The spectral region comprising the Xe5p and Xe5s emission is scaled by ×15. A NIR field-free photoelectron spectrum is depicted on the right hand side. (b) The COE analysis indicates that the emission of electrons liberated from the Xe5p orbitals is delayed by $\Delta\tau \approx 30$ asec with respect to photoelectrons released from the Xe4d subshell. The energy interval for the calculation of the first moments for the Xe5p streaking trace was chosen in a way to minimize the contribution of the weak Xe5s photoemission line. The kinetic energy is referenced to the vacuum level.

Even though this absolute time delay can only be deduced by theory, a strong experimental evidence for its existence has recently been established in streaking experiments performed on neon atoms in the gas phase. These measurements revealed a relative time delay of $\Delta\tau = 21 \pm 5$ asec between electrons released from the Ne2p orbitals and those liberated from the Ne2s orbital [44]. This relative delay in photoemission from an isolated atom cannot be explained by a simple propagation effect due to screening of the NIR streaking field by the remaining bound electrons surrounding the nucleus. Quite to the contrary, such a reasoning implies an earlier exposure of the faster Ne2p electrons to the streaking field and consequently a streaking time shift corresponding to a delayed emission of the slower Ne2s electrons, which is in disagreement with the experimental result. Whereas extensive theoretical modeling confirms the existence of this temporal effect in atomic photoemission, the exact value of the delay could so far not be reproduced. Even when electron-electron correlation was included, these calculations could only account for a delay of less than 10 asec [44, 169]. As proposed recently, the remaining discrepancy may be partly attributed to the influence of the NIR streaking field on the formation of the outgoing electron wave packets, i.e. to the breakdown of the strong-field approximation for slow electrons moving in a long-range potential of the ion. [169, 170, 128, 171, 172].

Obviously, such intrinsic time delays are also likely to occur in photoemission from the electronic states of xenon. This would further compromise the accuracy of the experimental approach pursued in the previous subsection, where it was intended to infer average photoelectron travel times in metals by using the photoemission from the Xe4d core level as a reference for prompt emission. In order to estimate the impact of this effect, additional streaking experiments have been performed on xenon atoms in the gas phase. A representative streaking spectrogram is depicted in Fig. 4.49 (a). For the gas-phase measurements, no background was subtracted from the electron spectra and the kinetic energy is given with respect to the vacuum level. The kinetic energy modulation revealing the waveform of the NIR vector potential is clearly resolved for both the Xe4d and Xe5p photoelectrons, whereas the interaction of the $N_{4,3}O_{2,3}O_{2,3}$ Auger electrons with NIR dressing field leads to the formation of sidebands. The latter is a consequence of the rather long duration of released Auger electron wave packets, which is governed by the Xe4d^{-1} core-hole lifetime of $\sim$7 fs [134].

A comparison of the two streaking traces extracted from the spectrogram is shown in Fig. 4.49 (b). It reveals a time shift of $\Delta\tau \approx 30$ between the electron release from the Xe4d and Xe5p orbitals, in reasonable agreement with the result of a TDSE-analysis of 7 streaking spectrograms (see Tab. 4.4). Further, several tests confirmed that this delay remains unaffected when the spectrogram is subjected to different background removal schemes. As in the case of neon, the faster photoelectrons (Xe5p) seem to lag behind the photoelectrons ejected with a lower kinetic energy, contrary to the intuitive expectation. First preliminary results from streaking measurements performed on xenon multilayers grown on W(110) further indicate that this temporal effect also persists for the condensed phase of xenon. For temperatures below 50 K, xenon can be grown epitaxially on a wide range of close-packed surfaces resulting in the formation of Xe(111) single crystals. A

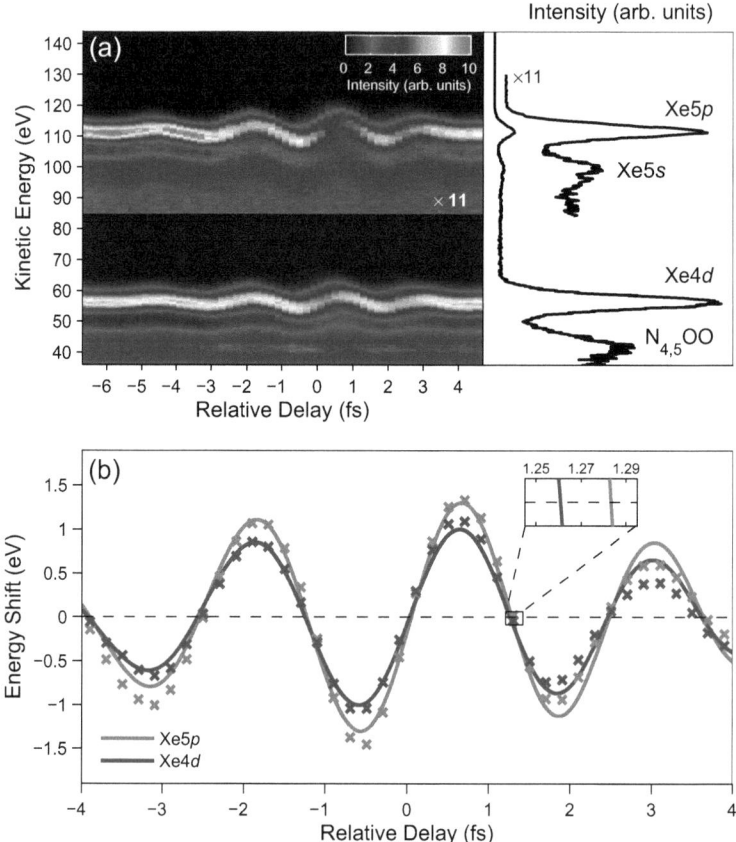

Figure 4.50: Attosecond streaking of a xenon multilayer condensed onto the W(110) crystal at 30 K. (a) Streaking spectrogram obtained with 118 eV XUV pulses. The spectral region comprising the Xe5p and Xe5s emission is scaled by ×11. An XUV-only photoelectron spectrum is shown on the right hand side. (b) The COE analysis suggest a small delay of $\Delta\tau \approx 20$ asec between the photoelectrons released from the Xe4p and Xe4d states. The first moments of the Xe5p streaking trace are evaluated in an energy interval chosen to eliminate the contribution of the weak Xe5s line

similar crystalline quality can therefore be assumed for the Xe multilayer condensed on the (110) surface of tungsten. A typical streaking spectrogram obtained from this system is shown in Fig. 4.50 (a). The sample was prepared *in situ* by backfilling the UHV chamber at a Xe partial pressure of $1 \cdot 10^{-8}$ mbar. The exposure was terminated after the W4f signal of the W(110) was completely obscured by the xenon overlayer, which sets a

Attosecond Streaking in Dielectrics: Xe/W(110)

Coverage	$\Delta\tau_{CB/5p}$ [asec]	Error($\pm$) [asec]	$\Delta\tau_{4f}$ [asec]	Error($\pm$) [asec]	Scans
1.0ML	13	10	95	10	7
2.0ML	30	8	150	8	2
3.5ML	15	9	188	9	1
$\geq$10ML	21	15	—	—	7
Xe gas	19	8	—	—	7

Table 4.4: Time delays of the CB/Xe5p electrons ($\Delta\tau_{CB/5p}$) and W4f electrons ($\Delta\tau_{4f}$) relative to the Xe4d emission for different Xe coverages on W(110). The error margins correspond to standard deviations. All time delays have been extracted from the streaking spectrograms with the TDSE-retrieval algorithm.

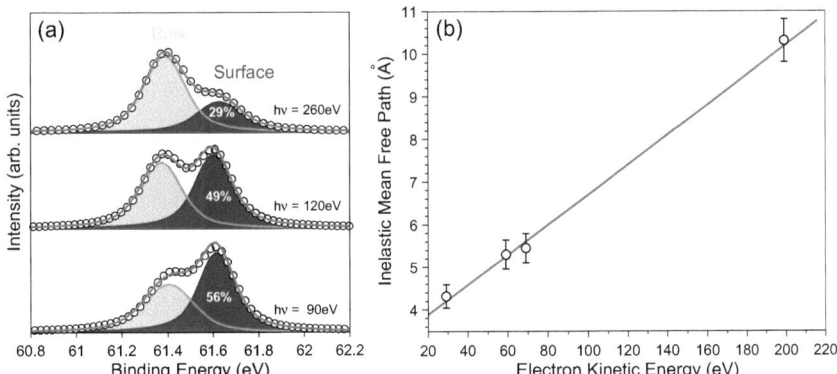

Figure 4.51: Analysis of the surface photoemission from solid Xenon. (a) Quantitative decomposition of the Xe4$d_{5/2}$ emission into surface and bulk components as a function photon energy. (b) Inelastic mean free path calculated from Eq. 4.7 with the interlayer spacing $d_{[111]} = 3.54$ Å of solid Xe(111) [163]. A linear fit to the data (red line) is used for interpolation.

lower limit for the final xenon coverage of $\Theta > 10\,\mathrm{ML}$.

The COE analysis shown in Fig. 4.50 (b) confirms a delay of the Xe5p emission of $\Delta\tau \approx 20$. It is worth mentioning that a time shift of similar magnitude can also be observed for all lower Xe coverages on W(110) investigated in this thesis (see Tab. 4.4). This insensitivity to the exact thickness of the xenon films corroborates the intrinsic nature of this relative time shift, with only negligible (or compensating) contributions related to condensed-matter-specific electron transport. This can be understood in terms of a simple compensating effect between the free-electron-like velocity and the average escape depth of the Xe5p and Xe4d photoelectrons leading to almost identical run-times for these electrons towards the xenon-vacuum interface.

For a quantitative estimate, the inelastic mean free path for electrons in solid xenon is needed which is not well-known in the kinetic energy range of interest for the experiments discussed here. However, similar to W(110) and Mg(0001), it may be inferred from the photon-energy-dependent branching into surface- and bulk-related components in high-resolution photoemission [173]. Exemplary spectra and fits to the Xe$4d_{5/2}$ photoemission obtained with narrow bandwidth synchrotron radiation from bulk xenon are shown in Figure 4.51 (a). The deconvolution has been performed with two doublets of Voigt functions over an energy range comprising the Xe$4d_{3/2}$ emission. Assuming the most likely arrangement of the xenon atoms in a Xe(111) crystal-like phase, the inelastic mean free path can be calculated from Eq. 4.7 with the interlayer spacing of bulk Xe(111) in [111]-direction of 3.54 Å [163]. The result is depicted in panel (b) together with a linear fit. As in the case of Mg(0001), the inelastic mean free path increases monotonically in the relevant range of electron energies which leads to a free-electron-like propagation time to the surface of 115 asec for *both* the slow Xe$4d$ electrons ($E_{kin} \approx 57$ eV) and the faster Xe$5p$ ($E_{kin} \approx 112$ eV) electrons. Evidently, this effect of cancellation between velocity and mean free path holds true for arbitrary NIR field strengths inside the xenon layers.

In summary, it must be concluded that absolute timing of photoemission from solids is not feasible without an improved understanding of the physical mechanisms underlying the intrinsic temporal structure observed in attosecond streaking of electron wave packets released from different energy levels of the same isolated atoms. It has to be emphasized that a measured *relative* time delay does not provide information on the time elapsed between the absorption of the XUV pulse and the ejection of the photoelectron, for any of the two orbitals involved. A precise knowledge of this time interval would be necessary for *all* electronic states in the system under scrutiny. Only then, a common time scale can be established on which temporal effects occurring in the release of electron wave packets from multiple energy levels and different atoms can be accurately compared and interpreted. For the two-electron system helium, these absolute time shift can indeed be retrieved from *ab initio* calculations [44]. Streaking experiments on a Xe-He gas mixture might therefore be helpful to establish an absolute reference time for the solid state Xe/W(110) systems. However, such experiments will suffer from the low cross-section for photo-ionization of the He$1s$ orbital. The feasibility of such measurements will therefore heavily depend on the progress made in increasing the XUV flux of high-harmonic sources.

Chapter 5

Summary, Conclusion & Outlook

The work presented in this thesis establishes laser-dressed attosecond photoemission as a quantitative technique to study ultrafast dynamics of photo-excited electrons at surfaces and interfaces with unprecedented temporal resolution. Within the pursued spectroscopic approach, photoelectron wave packets are launched inside the solid by the absorption of extreme ultraviolet (XUV) light bursts which last only for a few hundred attoseconds. The subsequent evolution and propagation of the generated photoelectrons is probed by their interaction with the electric field of a waveform-controlled near infrared (NIR) laser pulse precisely synchronized to the attosecond excitation. The strong electric field of the NIR pulse induces characteristic modulations in the kinetic energy spectrum of the XUV-ejected photoelectrons ("streaking"), which sensitively depend on the instant of their release into the laser field. This attosecond streaking method, which has hitherto been limited mainly to the study of isolated atoms in the gas phase, is successfully extended to various prototypical surface science model systems, encompassing atomically clean single crystalline metal surfaces, chemisorbed and physisorbed adsorbate layers as well as epitaxially grown metal-metal interfaces and rare-gas solids. The study of such a vast range of different systems is enabled by a newly developed apparatus combining state-of-the-art attosecond technology with established surface science techniques for sample preparation, handling and characterization.

Attosecond streaking experiments performed on tungsten single crystals confirm the existence of a time lag in the photoemission of core-level electrons compared to electrons released from the conduction band states of the solid, in agreement with the first proof-of-principle experiment reported by Cavalieri *et al.* in 2007. Beyond this basic verification, the refined measurements presented in this thesis approach an absolute accuracy of 10 asec and reveal that this relative time delay is only ∼30 asec for atomically clean surfaces but tends to increase up to ∼100 asec in the presence of impurity atoms on the surface. This clearly underlines the importance of maintaining excellent ultra-high vacuum conditions during the experiments in order to allow unambiguous quantitative interpretation of these subtle temporal effects. Further, no dependence of these time shifts on the strength of the applied NIR probing field could be detected for intensities below the damage threshold

Chapter 5. Summary, Conclusion & Outlook

of the solid surface, thereby demonstrating the robustness of the attosecond streaking technique for exploring dynamics in condensed matter on an attosecond time scale.

In contrast to the transition metal tungsten, an almost perfect synchronism between the release of photoelectrons from the core-level and the conduction band states is found for single crystal surfaces of the simple metal magnesium. Despite their rather different electronic structure, the time delays measured for both materials is shown to be compatible with a classical model in which photoelectrons emerge from different depths of the solid determined by their kinetic energy-dependent probability for inelastic scattering, and propagate with free-electron-like velocities towards the surface of the solid where they start interacting with the NIR streaking field. Within this model, these small time shifts arise as a general consequence of the monotonically growing escape depth with increasing kinetic energy of the electrons. A dominant influence on these time delays related to the different degrees of spatial localization of the involved initial-state wave functions or a possible modification of the photoelectron velocities due to the final-state band structure, which have recently been argued to be the main sources of these temporal effects, seems to be rather unlikely in the view of the experimental results obtained from these two systems.

In a second class of experiments, streaking spectroscopy is applied for the first time to surfaces covered by a well-defined number of epitaxial layers. This flexible sample configuration allows tracking the propagation of substrate core-level photoelectron wave packets through the overlayer material on the natural time scale of the electronic motion. Apart from demonstrating the capability of the streaking method to observe atomic-scale electron transport in real time, these measurements further substantiate the conclusion drawn from the streaking experiments on single crystal surfaces concerning the role of band structure for the propagation of photo-excited electrons inside the solid.

The analysis of streaked attosecond photoemission originating from metal surfaces covered by strongly interacting adsorbate monolayers of oxygen and magnesium reveals interesting temporal effects. For the magnesium monolayer, this effect manifests itself in an unexpectedly short time delay of the conduction band photoemission with respect to the emission from the core levels of the adsorbate layer. This phenomenon can be qualitatively understood in terms of strong hybridization between substrate band states and the valence levels of the adsorbate which gives rise to joint conduction band states with their associated wave functions being significantly delocalized across the substrate-adsorbate interface. Strong support for this interpretation is provided by complementary high-resolution synchrotron photoemission studies on these systems. A similar mechanism is proposed for the attosecond time-resolved photoemission from chemisorbed oxygen layers. Here, the valence charge redistribution at the metal-adsorbate interface induced by the hybridization alters the spatial onset of the screening of the NIR streaking field in the surface near region of the solid, which is detected in the experiment as an increased time delay between photoelectrons released from the substrate core-level states and the joint conduction band.

Chapter 5. Summary, Conclusion & Outlook

Measuring relative time delays occurring in the photoemission from different energy levels in a crystal does not yield any information on the absolute duration of the photo-ionization process in the solid, which might be defined as the time elapsed between the absorption of the attosecond XUV pulse and the appearance of the photoelectron in vacuum. These absolute durations might allow more detailed insight into the dynamics of many-body interactions accompanying the photoemission process, which are likely to cancel each other in relative timing experiments. The experimental approach pursued in this thesis to gain access to these absolute quantities draws on the concept of using photoemission from a weakly interacting adsorbate as a reference for clocking the photoelectron emission from the underlying metal substrate. In these studies, the screening of the NIR streaking field in the overlayer should be minimized in order to eliminate residual transport effects for photoelectrons within the adlayer, which makes dielectric adsorbates the prime candidates for providing this reference signal. First streaking experiments along these lines have been performed on tungsten surfaces covered with xenon atoms. It is demonstrated that streaking spectroscopy is applicable even to such weakly bound adsorbate systems and allows the acquisition of high-quality spectrograms without disturbance due to laser-induced desorption. The observed time shifts between substrate and adlayer photoemission are found to scale with the overlayer coverage, indicating that transport through the dielectric xenon adlayers is not negligible – even in the monolayer regime. Whereas this obstacle can in principle be overcome in future experiments by reducing the adsorbate coverage, a more fundamental limitation of this reference approach has been identified in the course of this thesis. Preliminary experiments performed on xenon in the gas phase reveal that time shifts of the order of ~ 20 asec are possible even between photoelectrons released from different orbitals of the xenon atom. Furthermore, this intrinsic time delay is shown to persist even for the condensed phase of xenon. The existence of such temporal effects for isolated atoms turns absolute time delays into theoretical constructs which can only be deduced from calculations, but remain inaccessible by attosecond streaking spectroscopy.

In conclusion, time-resolved attosecond photoemission from solids and atoms provides an interesting and wide field for further experimental and theoretical investigation. While no compelling evidence for band structure effects could be found in any experiments performed in this thesis, this aspect surely deserves further systematic examination. Streaking experiments performed with different XUV excitation energies on different crystal surfaces of the same material can help to finally clarify the importance of final-state effects. These experiments should be guided by detailed electronic structure calculation to selectively launch electron wave packets near band gaps or critical points of the material's band structure. For identifying the impact of Bloch group velocities on relative time shifts between electron wave packets released from a solid, it would be beneficial to employ higher XUV excitation energies, for which the simple free-electron propagation model predicts only negligible time delays for almost all materials. In addition, further work to determine the exact magnitude and the mechanism behind streaking time delays in atomic gases is highly necessary, since their understanding also forms the basis for a quantitative explanation of many temporal effects observed in photoemission from solids.

Chapter 5. Summary, Conclusion & Outlook

The results obtained in this thesis can therefore serve as a benchmark for a quantitative testing of theoretical approaches aiming to describe time-resolved attosecond photoemission in condensed matter.

The streaking experiments presented in this thesis suggest that time delays in solid-state photoemission are largely governed by transport phenomena which can be sufficiently accounted for by assuming free-electron-like dispersion. This will be valuable for the interpretation of future streaking experiments concentrating, for instance, on complex photoemission satellites, e.g. due to non-local screening (nickel oxide), where the measured time shifts can be expected to contain additional information on the dynamical electronic rearrangement in the vicinity of the photo-hole. The unique potential of the streaking technique to retrieve the full temporal evolution of the released electron wave packets (and not only the relative time delay in their propagation) may be exploited in experiments where attosecond pulses resonantly excite core-to-bound transitions. If this transition is triggered in an adsorbate strongly coupled to a surface, the electron wave packets released upon the decay of the resonance will reveal the complete time evolution of the excited electron as it delocalizes into the conduction band of the substrate. However, most of the prospects offered by attosecond streaking to shed new light on ultrafast electron dynamics are inextricably linked to the progress made in the generation of intense attosecond pulses at higher photon energies.

Appendix A

Calibration of Time-of-Flight Data

Attosecond photoemission experiments suffer from the low repetition rates of 1-4 kHz supported by current high-power laser amplifiers and the low conversion efficiency of high-harmonic generation for photon energies exceeding 100 eV. Since the resolution of the streaking technique is inextricably linked to the signal-to-noise ratio in the photoelectron spectra, the detection efficiency has to be maximized. In this respect, time-of-flight (TOF) spectrometers are superior to energy-dispersive analyzers since they allow, in principle, the collection of all the electrons produced by a single excitation pulse. In TOF detection, the kinetic energy E_{kin} of the electrons with respect to the Fermi level of the sample can be inferred from the measured electron flight times t according to[1]:

$$E_{kin} = \phi_A + \frac{1}{2} m_e \frac{L^2}{t^2}, \tag{A.1}$$

where ϕ_A is the work function and L the total drift length of the TOF spectrometer. The variable transformation from $t \to E_{kin}$ entails an additional rescaling of the intensities in the measured electron distribution N:

$$N(E_{kin}) = N(t) \left| \frac{dt}{dE_{kin}} \right| = \frac{t^3}{m_e L^2}. \tag{A.2}$$

The energy resolution ΔE_{kin} of the TOF spectrometer is given by:

$$\Delta E_{kin} = \sqrt{\frac{2}{m_e}} \frac{2}{L} \Delta t \, E_{kin}^{\ 3/2}, \tag{A.3}$$

where Δt denotes the minimal time interval that can be discriminated by the detection electronics. The commercial TOF spectrometer (Stefan Kaesdorf, Geräte für Forschung und Industrie) employed in all attosecond photoemission experiments reported in this thesis is schematically shown in Fig. A.1. It consists of a μ-metal-shielded field-free

---
[1] The acceleration of the electrons between the sample and the entrance of the spectrometer due to their different work functions can be safely neglected for kinetic energies > 10 eV.

Appendix A. Calibration of Time-of-Flight Data

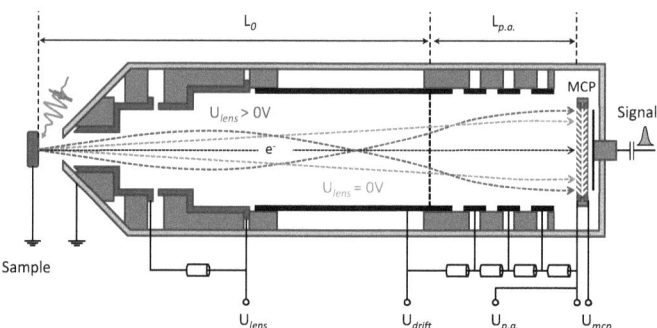

Figure A.1: Schematic of the employed time-of-flight (TOF) spectrometer. Depending on the (positive) voltage U_{lens} applied to the electrodes of the electrostatic lens, the collection efficiency can be enhanced for electrons within a certain kinetic energy range by increasing the effective acceptance angle up to $\pm 22°$. The regular acceptance angle for $U_{lens} = 0$ is $\approx \pm 2°$ (determined by the diameter of the MCPs and the length of the drift tube).

drift tube of length $L_0 = 425$ mm followed by a $L_{p.a.} = 70$ mm long segmented post-acceleration section in which the electrons are accelerated by a potential difference of $U_{p.a.} = 1000$ V onto the MCP detector. The resulting voltage pulses transmitted by a decoupling capacitor are amplified in a home-built pre-amplifier (5 GHz/ 40 dB) and are further processed by a constant fraction discriminator (ORTEC 9307) before being digitalized by a multi-scaler card with a time resolution of 100 ps (FAST ComTec P7889). The constant fraction discriminator ensures that the timing between the electron arrival at the detector and the external trigger signal (obtained from a photodiode in the laser system) is independent of the amplitude of the generated output pulses.

The spectrometer features an electrostatic lens which allows to increase the maximal detection angle for a limited range of electron energies, depending on the voltage U_{lens} applied to the electrodes. Because of the potentials U_{lens} and $U_{p.a.}$, the function relating t and E_{kin} is more complicated than suggested in Eq. A.1 and cannot be solved analytically anymore. For the necessary $t \to E_{kin}$ conversion, the relation between t and E_{kin} was therefore derived from electron trajectory simulations as a function of U_{lens} provided by the manufacturer. The corresponding intensity scaling coefficients $|dt/dE_{kin}|$ were obtained by differentiation of the simulated numerical data.

The kinetic energy-dependent transmission function introduced by the TOF lens was determined experimentally by comparing XUV-only spectra recorded with different settings of the lens voltage to a measurement performed without lens (see Fig. A.2 (b)). Sputtered Si(111) samples were used for this purpose since they feature a relatively smooth, yet sufficiently intense electron distribution (mainly due to the $L_{2,3}$VV Auger decay). A small negative bias voltage was applied to the sample to accurately measure the transmission for energies beyond the employed XUV photon energy. An example for the

Appendix A. Calibration of Time-of-Flight Data

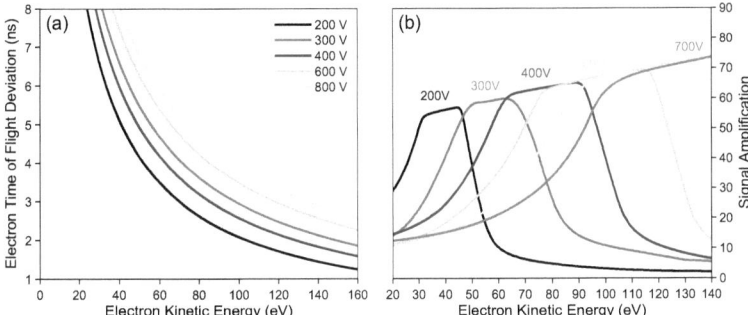

Figure A.2: Functions necessary for a complete calibration of raw TOF data. (a) Deviation of the electron flight time due to the voltage U_{lens} applied to the electrostatic lens. (b) Measured energy-dependence of the lens-induced signal amplification (after cubic spline interpolation).

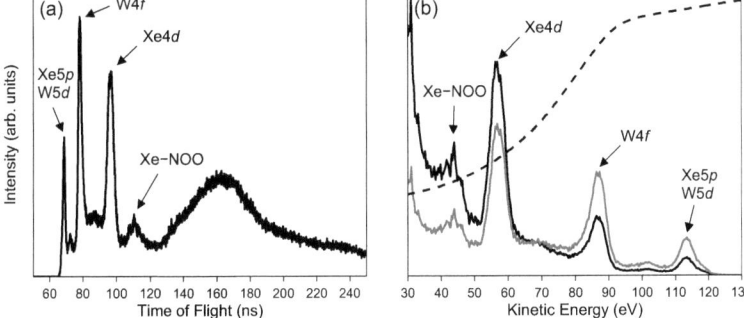

Figure A.3: Example of TOF data calibration. (a) Raw TOF electron spectrum of 2 ML Xe/W(110) recorded with a lens voltage of 600 V. (b) Spectrum after conversion to kinetic energy (gray) and after correction for the energy-dependent transmission of the electrostatic lens (black). The dashed line represents the transmission function corresponding to the applied lens voltage. TOF spectra calibrated in this way are in good agreement with synchrotron photoemission data in terms of relative intensities and peak positions.

complete calibration of raw time-of-flight data is shown in Fig. A.3 for a photoelectron spectrum acquired from 2 ML Xe/W(110). The work function of the analyzer was found to be $\phi_A \approx 4.5$ eV by comparison with calibrated synchrotron photoemission data. The lower limit for the energy resolution of the TOF spectrometer estimated from Eq. A.3 is $\Delta E_{kin} = 0.1 - 0.4$ eV for kinetic energies in the range of $50 - 130$ eV. However, both the timing jitter of the external trigger signal, and the averaging of time-of-flight signals due to the unresolved transverse momentum of the detected electrons are expected to further reduce the resolving power.

Appendix A. Calibration of Time-of-Flight Data

Appendix B

Accuracy of Delay-Extraction Procedures

The extraction of precise relative timing information from a measured streaking spectrogram is delicate. In general, any experimental deficiency that results in different distortions for different parts of the same spectrogram may lead to an erroneous retrieval of the time shifts encoded in the streaking measurement. It is therefore indispensable to test the robustness of the applied fitting schemes against such potential sources of systematic errors. Several influencing experimental conditions belonging to this category were identified in the course of this thesis, and their impact on the extracted time shifts was further studied in simulations. In the following, only the most important results and implications of this analysis are briefly discussed.

The selective signal amplification of the electrostatic TOF lens gives rise to a kinetic energy-dependent distortion in the photoelectron spectra. To study the effect of this spectral reshaping on the retrieved time delays, synthetical streaking spectrograms composed of two photoemission lines *without* any relative time delay in their emission were calculated (based on the TDSE-retrieval algorithm) with realistic parameters for the XUV and NIR pulses. Subsequently, the dependence of the retrieved time shift on different initial parameters and condition was systematically tested. An important result of these simulations is summarized in Fig. B.1. In cases where the transmission function of the TOF lens exhibits a large gradient in the spectral region comprising one of the photoemission lines, both the TDSE-retrieval and the simpler center-of-energy (COE) method retrieve an artificial delay when the electron wave packets are chirped. Moreover, this apparent time shift is strictly correlated with the sign and magnitude of the wave packet's chirp. In order to eliminate this effect, and to guarantee the highest accuracy for the extracted time delays presented in this thesis, *all* streaking spectrograms have been corrected and evaluated taking the measured transmission function into account. It should be emphasized that this systematic deficiency of the TDSE-fitting scheme in retrieving the true time shifts in distorted spectrograms could not even be overcome by constraining the

Appendix B. Accuracy of Delay-Extraction Procedures

wave packet durations and chirps to the correct initial values. Consequently, there is no alternative to an exact calibration of the streaking data, especially when small temporal effect are to be investigated.

The relevance of this phenomenon is demonstrated in Fig. B.2, taking the streaking time shift in gas phase neon as an experimental benchmark. For this system, a relative time delay of $\Delta\tau = 22 \pm 5$ asec between the photoemission of the Ne$2p$ and the Ne$2s$ electrons was recently measured, which represents the shortest time interval ever captured directly in a time-resolved experiment to date [44]. However, the analysis of 6 neon streaking spectrograms collected with lens voltages between $U_{lens} = 400 - 390$ V (see Fig. B.2 (c))

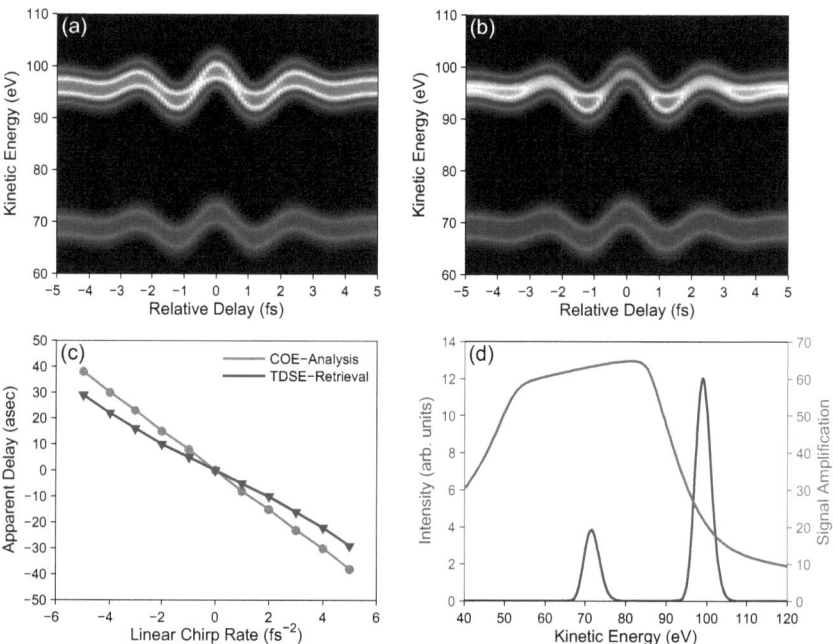

Figure B.1: Simulation of the systematic error introduced by the energy-dependent signal enhancement of the electrostatic TOF lens with realistic parameters: $I_L = 3 \cdot 10^{11}$ W/cm^2, $\tau_L = 5$ fs, $\lambda_L = 750$ nm, $\tau_x = 435$ asec. (a) Simulated spectrogram of two electron wave packets perfectly synchronized in time. (b) Same spectrograms but distorted by the selective signal amplification of the TOF lens ($U_{lens} = 390$ V). (c) Dependence of the time shift extracted from distorted spectrograms as a function of the wave packet's (linear) chirp b_x. Both the COE method and the TDSE-retrieval pretend the existence of a time shift between the two streaking traces which scales with the chirp carried by the electron wave packets. (d) Variation of the signal amplification in the spectral region containing the two photoemission lines.

Appendix B. Accuracy of Delay-Extraction Procedures

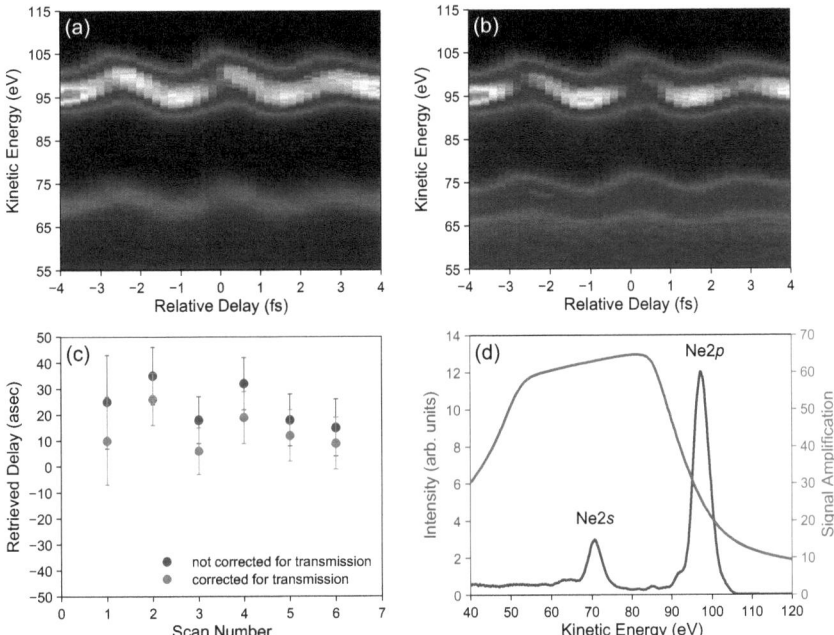

Figure B.2: Effects induced by the TOF lens in attosecond streaking experiments of gas-phase neon. Sub-fs XUV pulses were filtered with the multilayer mirror centered at ∼118 eV (4.2 eV FWHM). The voltage of the TOF lens was set to $U_{lens} = 440 - 390$ V to enhance the count rate of the weaker Ne2s line. (a) Measured spectrogram corrected for the transmission of the TOF lens. (b) Same spectrogram but without transmission correction. (c) Relative delay between the Ne2p and Ne2s photoemission extracted from a series of measurements with the TDSE-retrieval. The average delay of the Ne2p photoelectrons is 14 ± 7 asec when the effect of the transmission function is taken into account, whereas it is 24 ± 8 asec when the data is not corrected. (d) A NIR field-free electron spectrum of gas-phase neon is compared to the spectral shape of the lens transmission function for $U_{lens} = 390$ V.

reveals that the average time delay of the Ne2p photoelectrons is only 14 ± 7 asec when the effect of the lens-induced signal amplification is accounted for. In contrast, a mean time shift of 24 ± 8 asec is deduced from the same spectrograms when *no* transmission correction is applied. According to the simulation in Fig. B.1 (c), this difference of 10 asec corresponds exactly to the systematic error expected for wave packet chirps of -2 fs^2, which is very similar to the chirp introduced by the 118 eV multilayer mirror employed in these measurements (comp. Section 3.2.2). The existing discrepancy with the time delay published in [44] is subject of current investigation. Nevertheless, it has to be emphasized that all streaking measurements reported in [44] were acquired with a TOF spectrometer featuring a similar electrostatic lens, but a correction for the energy-

Appendix B. Accuracy of Delay-Extraction Procedures

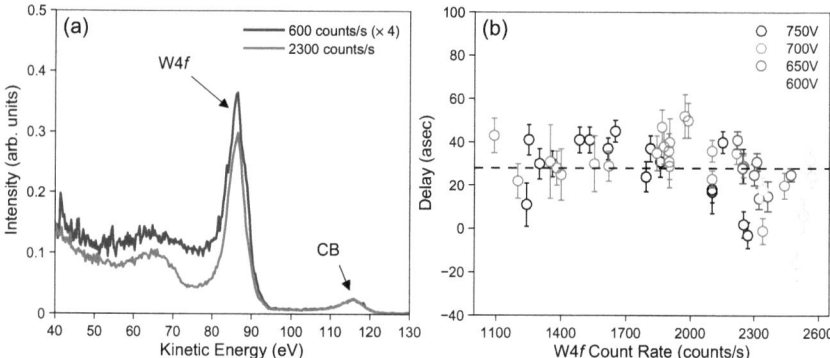

Figure B.3: Influence of TOF detector saturation on the extracted delay. (a) Photoelectron spectra of W(110) recorded with the detector driven into saturation (gray line) and well below the saturation level (dark gray line). For this purpose, the XUV flux was tuned by adjusting the Ne pressure in the HH target. (b) Time shift between CB and W4f in clean W(110) as a function of the average W4f count rate in the respective spectrograms. The dashed line represents the mean value for the delay of 28 asec.

dependent transmission was not performed for any of these data sets (comp. supporting information of [44]).

Towards the end of this thesis, a solution to circumvent the problems caused by the limited amplification bandwidth of the TOF lens was devised and successfully tested. By switching the lens voltage and the memory card of the time-to-digital converter in sync with the laser repetition rate, a simultaneous acquisition of two spectrograms becomes possible. The voltages can then be chosen independently for the two spectrograms to precisely focus on the photoemission lines of interest. This enhanced flexibility will contribute to expand the applicability of the attosecond streaking technique to systems with smaller photo-ionization cross-sections.

Another source of spectral distortions arises from signal saturation effects due to the electronic deadtime of the MCPs in TOF-spectrometer and the constant fraction discriminator. Due to the higher target density of the solid-state samples and the almost two order of magnitude signal amplification provided by the TOF lens, count rates approaching the repetition rate of 3 kHz were frequently achieved for the W4f emission of the clean W(110) surface. Under these conditions, saturation does not only lower the W4f-to-conduction band intensity ratio, but also leads to a slight narrowing of the W4f line-shape, and even modulates the electron background at lower kinetic energies (see Fig. B.3 (a)). For the streaking experiments on W(110) presented in the main text, lens voltages between 600 – 800 eV were chosen as a trade-off between enhancing the count rate in the CB region, and limiting at the same time the effect of saturation in the more intense W4f signal. However, even for these settings, saturation of the W4f peak could

Appendix B. Accuracy of Delay-Extraction Procedures

not be completely avoided. To check for a residual influence of these effects, all time shifts extracted from the W(110) spectrograms are plotted in Fig. B.3 (b) as a function of both the mean W4f count rate and the used TOF lens voltage. Apparently, a correlation might only be established between the retrieved time shift and the average W4f intensity for count rates exceeding 2200 counts/s. However, even if all measurements above this threshold are discarded, the mean delay is only 1 asec smaller than the mean value of 28 asec inferred from the full distribution. The influence of saturation on the time shifts presented in the main text is therefore negligible.

It is evident that the above-mentioned distortions of the spectrogram also affect the accuracy for retrieving other important electron wave packet properties like chirp and duration. Compared to the evaluation of relative time shifts, a reliable extraction of these parameters from the experimental data is more susceptible to imperfections of the measurement. For example, given the rather long acquisition time necessary for the photoelectron spectra comprising a spectrogram, any timing jitter between the XUV and NIR pulses (e.g. induced by mechanical vibrations coupling to the double mirror assembly) would translate into a broadening of the spectral features along the kinetic energy and the pump-probe delay axis. In unfavorable cases, this leads to an erroneous retrieval of the wave packet duration and chirp. In contrast, the relative time delay between two wave packets will hardly be affected since this distortion occurs equally for both parts of the streaking spectrogram. The same arguments apply to residual fluctuations and drifts of the carrier-envelope phase.

Appendix B. Accuracy of Delay-Extraction Procedures

Appendix C

Time Shifts & Spectral Resolution

The large bandwidth of the sub-fs XUV pulses requires electronic states that are well separated in energy to perform relative timing measurements by attosecond streaking, which is a severe limitation intrinsic to this method. It is therefore interesting to explore the possibility to what extent reliable timing information can be retrieved from a streaked photoelectron distribution of spectrally not fully resolved energy levels. In this respect, the overlapping W5d and Xe5p states of 1 ML xenon on W(110) may serve as a test case. For sub-fs XUV photo-excitation, the valence region of this system has a characteristic asymmetric line-shape which can be described as an incoherent superposition of the CB emission from clean W(110) and broad feature containing the Xe5p doublet (see Fig. C.1).

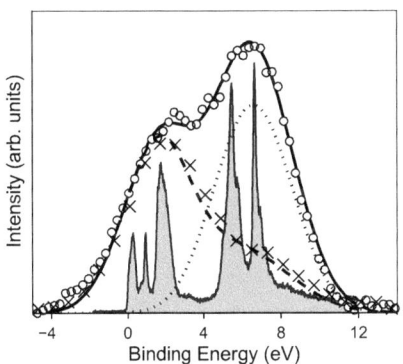

Figure C.1:
Decomposition of the valence emission from 1 ML Xe/W(110) into contributions due to the clean W(110) surface and the atomic-like Xe5p states. A spectrum obtained with the 118 eV XUV mirror from clean W(110) is shown crosses. Circles denote a corresponding spectrum acquired from 1 ML Xe/W(110). The corresponding model electron spectra are shown as solid and dashed line. Their difference (dotted line) approximates the isolated contribution of the Xe5p states which is used in the simulations shown in Fig. C.2. A synchrotron spectrum of 1 ML Xe/W(110) is shown as a reference (gray shaded area).

The simulations shown in Fig. C.2 demonstrate that a time delay in the electron emission from these two different states will manifest itself in a subtle modulation of their overlapping electron distributions as a function of the NIR-XUV delay. Further tests reveal

Appendix C. Time Shifts & Spectral Resolution

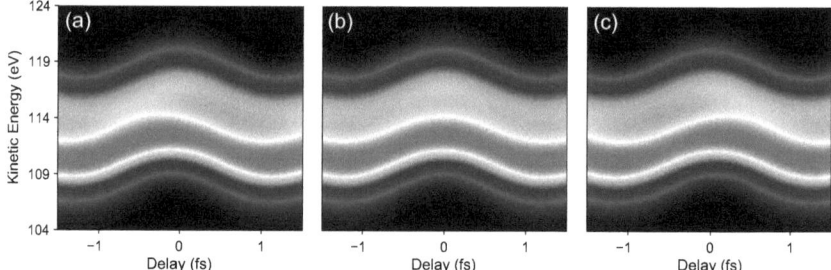

Figure C.2: Simulated streaking spectrograms of the overlapping W5d and Xe5p valence states in 1 ML Xe/W(110) based on the decomposition depicted in Fig. C.1. The exciting Gaussian XUV pulse is transform-limited with a duration of $\tau_x = 435$ asec. Spectrograms are constructed for (a) a delayed emission of the Xe5p states by 100 ascec, (b) perfectly synchronized emission of W5d and Xe5p electrons, and (c) a delayed emission of the W5d electrons by 100 asec. The sign of the relative time shift can be clearly inferred from the spectrograms.

that time shifts down to 10 asec can be reliably extracted with the TDSE-retrieval from such spectrograms, even for realistic signal-to-noise ratios. Unfortunately, these characteristic modulations become gradually obscured for chirped electron wave packets, which rendered a corresponding analysis of the experimental data obtained in this thesis ambiguous. Nevertheless, the performed simulations indicate that such a time shift between spectrally overlapping states can be resolved in future streaking experiments when nearly Fourier-limited XUV pulses are employed. However, a precise knowledge of the system's electronic structure, e.g. from synchrotron studies, is obviously a prerequisite for such an advanced analysis.

Bibliography

[1] A. H. Zewail, *J. Phys. Chem. A* **104**, 5660 (2000).

[2] F. Krausz and M. Ivanov, *Rev. Mod. Phys.* **81**, 163 (2009).

[3] T. Brabec and F. Krausz, *Rev. Mod. Phys.* **72**, 545 (2000).

[4] A. Baltuška, Th. Udem, M. Uiberacker, M. Hentschel, E. Goulielmakis, Ch. Gohle, R. Holzwarth, V. S. Yakovlev, A. Scrinzi, T. W. Hänsch, and F. Krausz, *Nature* **421**, 611 (2003).

[5] P. M. Paul, E. S. Toma, P. Breger, G. Mullot, F. Auge, Ph. Balcou, H. G. Müller, and P. Agostini, *Science* **292**, 1689 (2001).

[6] M. Drescher, M. Hentschel, R. Kienberger, G. Tempea, C. Spielmann, G. A. Reider, P. B. Corkum, and F. Krausz, *Science* **291**, 1923 (2001).

[7] M. Hentschel, R. Kienberger, Ch. Spielmann, G. A. Reider, N. Milosevic, T. Brabec, P. Corkum, U. Heinzmann, M. Drescher, and F. Krausz, *Nature* **414**, 509 (2001).

[8] I. P. Christov, M. M. Murnane, and H. C. Kapteyn, *Phys. Rev. Lett.* **78**, 1251 (1997).

[9] E. Goulielmakis, M. Schultze, M. Hofstetter, V. S. Yakovlev, J. Gagnon, M. Uiberacker, A. L. Aquila, E. M. Gullikson, D. T. Attwood, R. Kienberger, F. Krausz, and U. Kleineberg, *Science* **320**, 1614 (2008).

[10] M. Uiberacker, Th. Uphues, M. Schultze, A. J. Verhoef, V. Yakovlev, M. F. Kling, J. Rauschenberger, N. M. Kabachnik, H. Schroder, M. Lezius, K. L. Kompa, H.-G. Muller, M. J. J. Vrakking, S. Hendel, U. Kleineberg, U. Heinzmann, M. Drescher, and F. Krausz, *Nature* **446**, 627 (2007).

[11] E. Goulielmakis, Z.-H. Loh, Adrian W., R. Santra, N. Rohringer, V. S. Yakovlev, S. Zherebtsov, T. Pfeifer, A. M. Azzeer, M. F. Kling, S. R. Leone, and F. Krausz, *Nature* **466**, 739 (2010).

[12] S. Hüfner, *Photoelectron Spectroscopy - Principles and Applications*, Springer-Verlag, (2003).

[13] W. Schattke and M. A. Van Hove, *Solid-State Photoemission and Related Methods: Theory and Experiment*, Wiley-VCH, (2003).

Bibliography

[14] R. Haight, *Surface Science Reports* **21**, 275 (1995).

[15] H. Petek and S. Ogawa, *Progr. in Surface Science* **56**, 239 (1997).

[16] M. Weinelt, *J. Phys. C* **14**, R1099 (2002).

[17] A. Borisov, D. Sánchez-Portal, R. Díez Muiño, and P. M. Echenique, *Chem. Phys. Lett.* **387**, 95 (2004).

[18] A. Föhlisch, P. Feulner, F. Hennies, A. Fink, D. Menzel, D. Sanchez-Portal, P. M. Echenique, and W. Wurth, *Nature* **436**, 373 (2005).

[19] A. G. Borisov, A. K. Kazansky, and J. P. Gauyacq, *Phys. Rev. Lett.* **80**, 1996 (1998).

[20] A. G. Borisov, A. K. Kazansky, and J. P. Gauyacq, *Phys. Rev. B* **59**, 10935 (1999).

[21] P. A. Brühwiler, O. Karis, and N. Mårtensson, *Rev. Mod. Phys.* **74**, 703 (2002).

[22] A. L. Cavalieri, N. Müller, Th. Uphues, V. S. Yakovlev, A. Baltuška, B. Horvath, B. Schmidt, L. Blümel, R. Holzwarth, S. Hendel, M. Drescher, U. Kleineberg, P. M. Echenique, R. Kienberger, F. Krausz, and U. Heinzmann, *Nature* **449**, 1029 (2007).

[23] J.-C. Diels and W. Rudolph, *Ultrashort Laser Pulse Phenomena*, Academic Press, (2006).

[24] C. Rullière, *Femtosecond Laser Pulses*, Springer-Verlag, (2004).

[25] H. Lüth, *Surfaces and Interfaces of Solids*, Springer-Verlag, (1993).

[26] A. Zangwill, *Physics at Surfaces*, Cambridge University Press, (1988).

[27] A. McPherson, G. Gibson, H. Jara, U. Johann, T. S. Luk, I. A. McIntyre, K. Boyer, and C. K. Rhodes, *J. Opt. Soc. Am. B* **4**, 595 (1987).

[28] P.B. Corkum, *Phys. Rev. Lett.* **71**, 1994 (1993).

[29] M. Lewenstein, Ph. Balcou, M. Yu. Ivanov, L'Huillier A, and P. B. Corkum, *Phy. Rev. A* **49**, 2117 (1994).

[30] Ch. Spielmann, N. H. Burnett, S. Sartania, R. Koppitsch, M. Schnurer, C. Kan, M. Lenzner, P. Wobrauschek, and F. Krausz, *Science* **278**, 661 (1997).

[31] J. Seres, E. Seres, A. J. Verhoef, G. Tempea, C. Streli, P. Wobrauschek, V. Yakovlev, A. Scrinzi, C. Spielmann, and F. Krausz, *Nature* **433**, 596 (2005).

[32] A. Wonisch, U. Neuhäusler, N. M. Kabachnik, T. Uphues, M. Uiberacker, V. Yakovlev, F. Krausz, M. Drescher, U. Kleineberg, and U. Heinzmann, *Appl. Opt.* **45**, 4147 (2006).

[33] M. Hofstetter, M. Schultze, M. Fieß, B. Dennhardt, A. Guggenmos, J. Gagnon, V. S. Yakovlev, E. Goulielmakis, R. Kienberger, E. M. Gullikson, F. Krausz, and U. Kleineberg, *Opt. Express* **19**, 1767 (2011).

[34] R. Kienberger, E. Goulielmakis, M. Uiberacker, A. Baltuška, V. Yakovlev, F. Bammer, A. Scrinzi, Th. Westerwalbesloh, U. Kleineberg, U. Heinzmann, M. Drescher, and F. Krausz, *Nature* **427**, 817 (2004).

Bibliography

[35] P. Salières, T. Ditmire, K. S. Budil, M. D. Perry, and A. L'Huillier, *J. Phys. B* **27**, L217 (1994).

[36] P. Salières, T. Ditmire, M. D. Perry, A. L'Huillier, and M. Lewenstein, *J. Phys. B* **29**, 4771 (1996).

[37] F. Linder, G. G. Paulus, H. Walther, A. Baltuška, E. Goulielmakis, M. Lezius, and F. Krausz, *Phys. Rev. Lett.* **92**, 113001 (2004).

[38] M. Schnürer, Z. Cheng, M. Hentschel, F. Krausz, T. Wilhein, D. Hambach, G. Schmahl, M. Drescher, Y. Lim, and U. Heinzmann, *Appl. Phys. B* **70**, S227 (2000).

[39] C. Altucci, J.W.G. Tisch, and R. Velotta, *Journal of Modern Optics* **58**, 1585 (2011).

[40] G. Sansone, E. Benedetti, F. Calegari, C. Vozzi, L. Avaldi, R. Flammini, L. Poletto, P. Villoresi, C. Altucci, R. Velotta, S. Stagira, S. De Silvestri, and M. Nisoli, *Science* **314**, 443 (2006).

[41] H. Mashiko, S. Gilbertson, C. Li, S. D. Khan, M. M. Shakya, E. Moon, and Z. Chang, *Phys. Rev. Lett.* **100**, 103906 (2008).

[42] M. Geissler, G. Tempea, and T. Brabec, *Phys. Rev. A* **62**, 033817 (2000).

[43] M. Drescher, M. Hentschel, R. Kienberger, M. Uiberacker, V. Yakovlev, A. Scrinzi, Th. Westerwalbesloh, U. Kleineberg, U. Heinzmann, and F. Krausz, *Nature* **419**, 803 (2002).

[44] M. Schultze, M. Fiess, N. Karpowicz, J. Gagnon, M. Korbman, M. Hofstetter, S. Neppl, A. L. Cavalieri, Y. Komninos, Th. Mercouris, C. A. Nicolaides, R. Pazourek, S. Nagele, J. Feist, J. Burgdörfer, A. M. Azzeer, R. Ernstorfer, R. Kienberger, U. Kleineberg, E. Goulielmakis, F. Krausz, and V. S. Yakovlev, *Science* **328**, 1658 (2010).

[45] J. Itatani, F. Quéré, G. L. Yudin, M. Yu. Ivanov, F. Krausz, and P. B. Corkum, *Phys. Rev. Lett.* **88**, 173903 (2002).

[46] M. Kitzler, N. Milosevic, A. Scrinzi, F. Krausz, and T. Brabec, *Phys. Rev. Lett.* **88**, 173904 (2002).

[47] E. Goulielmakis, M. Uiberacker, R. Kienberger, A. Baltuška, V. Yakovlev, A. Scrinzi, Th. Westerwalbesloh, U. Kleineberg, U. Heinzmann, M. Drescher, and F. Krausz, *Science* **305**, 1267 (2004).

[48] F. Quéré, Y. Mairesse, and J. Itatani, *J. Mod. Opt.* **52**, 339 (2005).

[49] Y. Mairesse and F. Quéré, *Phys. Rev. A* **71**, 011401 (2005).

[50] J. Gagnon, E. Goulielmakis, and V.S. Yakovlev, *Appl. Phys. B* **92**, 25 (2008).

[51] D. M. Volkow, *Zeitschrift für Physik A* **94**, 250 (1935).

[52] V. S. Yakovlev, J. Gagnon, N. Karpowicz, and F. Krausz, *Phys. Rev. Lett.* **105**, 073001 (2010).

Bibliography

[53] M. Wickenhauser, J. Burgdörfer, F. Krausz, and M. Drescher, *Phys. Rev. Lett.* **94**, 023002 (2005).

[54] L. Miaja-Avila, G. Saathoff, S. Mathias, J. Yin, C. La o vorakiat, M. Bauer, M. Aeschlimann, M. M. Murnane, and H. C. Kapteyn, *Phys. Rev. Let.* **101**(4), 046101 (2008).

[55] C. N. Berglund and W. E. Spicer, *Phys. Rev.* **136**, A1030 (1964).

[56] G. D. Mahan, *Phys. Rev. B* **2**, 4334 (1970).

[57] E. E. Krasovskii and M. Bonitz, *Phys. Rev. Lett.* **99**, 247601 (2007).

[58] L. Miaja-Avila, C. Lei, M. Aeschlimann, J. L. Gland, M. M. Murnane, H. C. Kapteyn, and G. Saathoff, *Phys. Rev. Lett.* **97**, 113604 (2006).

[59] A. K. Kazansky and P. M. Echenique, *Phys. Rev. Lett.* **102**, 177401 (2009).

[60] E. E. Krasovskii, *Phys. Rev. B* **84**, 195106 (2011).

[61] C.-H. Zhang and U. Thumm, *Phys. Rev. Lett.* **102**, 123601 (2009).

[62] C.-H. Zhang and U. Thumm, *Phys. Rev. Lett.* **103**, 239902 (2009).

[63] C. Lemell, B. Solleder, K. Tőkési, and J. Burgdörfer, *Phy. Rev. A* **79**, 062901 (2009).

[64] P. J. Feibelman, *Phys. Rev. B* **12**, 1319 (1975).

[65] M. Schultze, A. Wirth, I. Grguras, M. Uiberacker, T. Uphues, A.J. Verhoef, J. Gagnon, M. Hofstetter, U. Kleineberg, E. Goulielmakis, and F. Krausz, *J. Electron Spectrosc. Relat. Phenom.* **184**, 68 (2011).

[66] M. Fieß, M. Schultze, E. Goulielmakis, B. Dennhardt, J. Gagnon, M. Hofstetter, R. Kienberger, and F. Krausz, *Rev. Sci. Instr.* **81**, 093103 (2010).

[67] M. J. Abel, Thomas Pfeifer, P. M. Nagel, W. Boutu, M. J. Bell, C. P. Steiner, Daniel M. Neumark, and S. R. Leone, *Chem. Phys.* **366**, 9 (2009).

[68] A. L. Cavalieri, E. Goulielmakis, B. Horvath, W. Helml, M. Schultze, M. Fieß, V. Pervak, L. Veisz, V. S. Yakovlev, M. Uiberacker, A. Apolonski, F. Krausz, and R. Kienberger, *New J. Phys.* **9**, 242 (2007).

[69] B. Horvárth, *Generation, characterization and sub-cycle shaping of intense, few-cycle light waveforms for attosecond spectroscopy*, PhD thesis Ludwig-Maximilians-Universität München (2011).

[70] R. Szipöcs, K. Ferencz, C. Spielmann, and F. Krausz, *Opt. Lett.* **19**, 201 (1994).

[71] A. Stingl, C. Spielmann, F. Krausz, and R. Szipöcs, *Opt. Lett.* **19**, 204 (1994).

[72] A. Stingl, M. Lenzner, Ch. Spielmann, F. Krausz, and R. Szipöcs, *Opt. Lett.* **20**, 602 (1995).

[73] D. Strickland and G. Mourou, *Optics Communications* **55**, 447 (1985).

[74] S. A. Planas, N. L. Pires Mansur, C. H. Brito Cruz, and H. L. Fragnito, *Opt. Lett.* **18**, 699 (1993).

[75] Zhenghu Chang, *Fundamentals of Attosecond Optics*, CRC Press, (2011).

[76] M. Nisoli, S. De Silvestri, O. Svelto, R. Szipöcs, K. Ferencz, Ch. Spielmann, S. Sartania, and F. Krausz, *Opt. Lett.* **22**, 522 (1997).

[77] S. T. Cundiff and J. Ye, *Rev. Mod. Phys.* **75**, 325 (2003).

[78] D. J. Jones, S. A. Diddams, J. K. Ranka, A. Stentz, R. S. Windeler, J. L. Hall, and S. T. Cundiff, *Science* **288**, 635 (2000).

[79] J. Reichert, R. Holzwarth, Th. Udem, and T.W. Hänsch, *Optics Communications* **172**, 59 (1999).

[80] R. Holzwarth, Th. Udem, T. W. Hänsch, J. C. Knight, W. J. Wadsworth, and P. St. J. Russell, *Phys. Rev. Lett.* **85**, 2264 (2000).

[81] T. Fuji, J. Rauschenberger, C. Gohle, A. Apolonski, T. Udem, V. S. Yakovlev, G. Tempea, T. W Hänsch, and F. Krausz, *New J. Phys.* **7**, 116 (2005).

[82] T. Fuji, J. Rauschenberger, A. Apolonski, V. S. Yakovlev, G. Tempea, T. Udem, C. Gohle, T. W. Hänsch, W. Lehnert, M. Scherer, and F. Krausz, *Opt. Lett.* **30**, 332 (2005).

[83] T. Wittmann, B. Horvath, W. Helml, M. G. Schätzel, X. Gu, A. L. Cavalieri, G. G. Paulus, and R. Kienberger, *Nature Physics* **5**, 357 (2009).

[84] G. G. Paulus, F. Grasbon, H. Walther, P. Villoresi, M. Nisoli, S. Stagira, E. Priori, and S. De Silvestri, *Nature* **414**, 182 (2001).

[85] E. Magerl, S. Neppl, A. L. Cavalieri, E. M. Bothshafter, M. Stanislawski, Th. Uphues, M. Hofstetter, U. Kleineberg, J. V. Barth, D. Menzel, F. Krausz, R. Ernstorfer, R. Kienberger, and P. Feulner, *Rev. Sci. Instr.* **82**, 063104 (2011).

[86] E. Magerl, *Attosecond photoelectron spectroscopy of electron transport in solids*, PhD Thesis Ludwig-Maximilians-Universität München (2011).

[87] G. Tempea, M. Geissler, M. Schnürer, and T. Brabec, *Phys. Rev. Lett.* **84**, 4329 (2000).

[88] Website of Berkeley Center for X-ray optics (2011), http://http://henke.lbl.gov/optical_constants/.

[89] R. Scheuerer, *Polarisations- und winkelaufgelöste Photodesorptionsmessungen aus molekularen Kondensaten nach Rumpfanregung. Zeitskalen im Desorptionsprozeß*, PhD Thesis Technische Universität München (1996).

[90] V. S. Yakovlev, F. Bammer, and A. Scrinzi, *J. Mod. Opt.* **52**, 305 (2005).

[91] Website of BESSY-II (2011), http://www.bessy.de/bit/upload/id_15_1.pdf.

[92] K. Wille, *The Physics of Particle Accelerators*, Oxford University Press, (2000).

Bibliography

[93] M. Krumrey, E. Tegeler, J. Barth, M. Kirsch, F. Schäfers, and R. Wolf, *Applied Optics* **27**, 4336 (1988).

[94] M. Schultze, E. Goulielmakis, M. Uiberacker, M. Hofstetter, J. Kim, D. Kim, F. Krausz, and U. Kleineberg, *New J. Phys.* **9**(7), 243 (2007).

[95] C.-H. Zhang and U. Thumm, *Phys. Rev. A* **84**, 033401 (2011).

[96] J. A. Becker, E. J. Becker, and R. G. Brandes, *J. Appl. Phys.* **32**, 411 (1961).

[97] M. Bode, S. Krause, L. Berbil-Bautista, S. Heinze, and R. Wiesendanger, *Surface Science* **601**, 3308 (2007).

[98] S. Doniach and M. Šunjić, *J. Phys. C* **3**, 285 (1970).

[99] D. M. Riffe, G. K. Wertheim, and P. H. Citrin, *Phys. Rev. Lett.* **63**, 1976 (1989).

[100] J. H. Weaver, C. G. Olson, and D. W. Lynch, *Phys. Rev. B* **12**, 1293 (1975).

[101] N. E. Christensen and B. Feuerbacher, *Phys. Rev. B* **10**, 2349 (1974).

[102] R. H. Gaylord and S. D. Kevan, *Phys. Rev. B* **36**, 9337 (1987).

[103] T.-W. Pi, L.-H. Hong, R.-T. Wu, and C.-P. Cheng, *Surface Review and Letters* **4**, 1197 (1997).

[104] R. F. Willis and N. E. Christensen, *Phys. Rev. B* **18**, 5140 (1978).

[105] J. C. Baggesen and L. B. Madsen, *Phys. Rev. A* **80**, 030901 (2009).

[106] S. Hellmann, K. Rossnagel, M. Marczynski-Bühlow, and L. Kipp, *Phys. Rev. B* **79**, 035402 (2009).

[107] G. Saathoff, L. Miaja-Avila, M. Aeschlimann, M. M. Murnane, and H. C. Kapteyn, *Phys. Rev. A* **77**, 022903 (2008).

[108] D. Briggs and M.P. Seah, *Practical Surface Analysis*, John Wiley & Sons, (1990).

[109] D.A. Shirley, *Phys. Rev.* **5**, 4709 (1972).

[110] M. Uiberacker, E. Goulielmakis, R. Kienberger, A. Baltuška, T. Westerwalbesloh, U. Kleineberg, U. Heinzmann, M. Drescher, and F. Krausz, *Laser Physics* **15**, 195 (2005).

[111] E. G. Gamaly, A. V. Rode, B. Luther-Davies, and V. T. Tikhonchuk, *Physics of Plasmas* **9**, 949 (2002).

[112] J.-J. Yeh, *Atomic Calculation of Photoionization Cross-Sections and Asymmetry Parameters*, Gordon and Breach Science Publishers, (1993).

[113] C. H.F. Peden and N. D. Shinn, *Surface Science* **312**, 151 (1994).

[114] R. A. Bartynski, R. H. Gaylord, T. Gustafsson, and E. W. Plummer, *Phys. Rev. B* **33**, 3644 (1986).

Bibliography

[115] F. Schiller, M. Heber, V. D. P. Servedio, and C. Laubschat, *Phys. Rev. B* **70**, 125106 (2004).

[116] R. Kammerer, J. Barth, F. Gerken, C. Kunz, S. A. Flodstrøm, and L. I. Johansson, *Phys. Rev. B* **26**, 3491 (1982).

[117] U. O. Karlsson, G. V. Hansson, P. E. S. Persson, and S. A. Flodström, *Phys. Rev. B* **26**, 1852 (1982).

[118] David R. Penn, *Phys. Rev. Lett.* **40**, 568 (1978).

[119] M. Aeschlimann, C. A. Schmuttenmaer, H. E. Elsayed-Ali, R. J. D. Miller, J. Cao, Y. Gao, and D. A. Mantell, *J. Chem. Phys.* **102**, 8606 (1995).

[120] M.A. Van Hove and S.Y. Tong, *Surface Crystallography by LEED*, Springer-Verlag, (1979).

[121] S. Tanuma, C. J. Powell, and D. R. Penn, *Surface and Interface Analysis* **43**, 689 (2011).

[122] L. I. Johansson and Bo E. Sernelius, *Phys. Rev. B* **50**, 16817 (1994).

[123] J. D. Bourke and C. T. Chantler, *Phys. Rev. Lett.* **104**, 206601 (2010).

[124] E. E. Krasovskii, V. M. Silkin, V. U. Nazarov, P. M. Echenique, and E. V. Chulkov, *Phys. Rev. B* **82**, 125102 (2010).

[125] E.D. Palik, *Handbook of Optical Constants of Solids*, Academic Press Inc, (1991).

[126] E.V. Chulkov, V.M. Silkin, and E.N. Shirykalov, *Surface Science* **188**, 287 (1987).

[127] L. Aballe, C. Rogero, and K. Horn, *Phys. Rev. B* **65**, 125319 (2002).

[128] K. Klünder, J. M. Dahlström, M. Gisselbrecht, T. Fordell, M. Swoboda, D. Guénot, P. Johnsson, J. Caillat, J. Mauritsson, A. Maquet, R. Taïeb, and A. L'Huillier, *Phys. Rev. Lett.* **106**, 143002 (2011).

[129] E. Taft and L. Apker, *Phys. Rev.* **99**, 1831 (1955).

[130] L. Ley, F. R. McFeely, S. P. Kowalczyk, J. G. Jenkin, and D. A. Shirley, *Phys. Rev. B* **11**, 600 (1975).

[131] R. Huber, F. Tauser, A. Brodschelm, M. Bichler, G. Abstreiter, and A. Leitenstorfer, *Nature* **414**, 286 (2001).

[132] D. Norman and D.P. Woodruff, *Surface Science* **79**, 76 (1979).

[133] J.C. Ashley, C.J. Tung, and R.H. Ritchie, *Surface Science* **81**, 409 (1979).

[134] J.L. Campbell and T. Papp, *Atomic Data and Nuclear Data Tables* **77**, 1 (2001).

[135] E. Umbach, J.C. Fuggle, and D. Menzel, *J. Electron Spectrosc. Relat. Phenom.* **10**, 15 (1977).

[136] L.H. Germer and J.W. May, *Surface Science* **4**, 452 (1966).

Bibliography

[137] R. X. Ynzunza, F. J. Palomares, E. D. Tober, Z. Wang, J. Morais, R. Denecke, H. Daimon, Y. Chen, Z. Hussain, M. A. Van Hove, and C. S. Fadley, *Surface Science* **442**, 27 (1999).

[138] H. Daimon, R. Ynzunza, J. Palomares, H. Takabi, and C. S. Fadley, *Surface Science* **408**, 260 (1998).

[139] K. E. Johnson, R. J. Wilson, and S. Chiang, *Phys. Rev. Lett.* **71**, 1055 (1993).

[140] K.J. Rawlings, *Surface Science* **99**, 507 (1980).

[141] D. M. Riffe and G. K. Wertheim, *Surface Science* **399**, 248 (1998).

[142] M. Stöhr, R. Podloucky, and S. Müller, *J. Phys. C.* **21**, 134017 (2009).

[143] J. Feydt, A. Elbe, H. Engelhard, G. Meister, and A. Goldmann, *Surface Science* **440**, 213 (1999).

[144] J. C. Baggesen and L. B. Madsen, *Phys. Rev. A* **78**, 032903 (2008).

[145] M. Stöhr, *First principles investigations of adsorption on a transition metal surface: oxygen and indium on the (110) surface of tungsten*, PhD Thesis Universität Wien (2009).

[146] N. Vinogradov, D. Marchenko, A. Shikin, V. Adamchuk, and O. Rader, *Physics of the Solid State* **51**, 179 (2009).

[147] A. Shikin, D. Marchenko, N. Vinogradov, G. Prudnikova, A. Rybkin, V. Adamchuk, and O. Rader, *Physics of the Solid State* **51**, 608 (2009).

[148] L. Aballe, A. Barinov, A. Locatelli, T. O. Mentes, and M. Kiskinova, *Phys. Rev. B* **75**, 115411 (2007).

[149] H. Over, T. Hertel, H. Bludau, S. Pflanz, and G. Ertl, *Phys. Rev. B* **48**, 5572 (1993).

[150] T.-W. Pi, I.-H. Hong, and C.-P. Cheng, *Phys. Rev. B* **58**, 4149 (1998).

[151] A. M. Shikin and O. Rader, *Phys. Rev. B* **76**, 073407 (2007).

[152] P. A. Thiry, J. Ghijsen, R. Sporken, J. J. Pireaux, R. L. Johnson, and R. Caudano, *Phys. Rev. B* **39**, 3620 (1989).

[153] L. Aballe, A. Barinov, A. Locatelli, S. Heun, and M. Kiskinova, *Phys. Rev. Lett.* **93**, 196103 (2004).

[154] N. Binggeli and M. Altarelli, *Phys. Rev. B* **78**, 035438 (2008).

[155] A.U. Goonewardene, J. Karunamuni, R.L. Kurtz, and R.L. Stockbauer, *Surface Science* **501**, 102 (2002).

[156] A. Hitachi, V. Chepel, M. I. Lopes, and V. N. Solovov, *J. Chem. Phys.* **123**, 234508 (2005).

[157] A. C. Sinnock and B. L. Smith, *Phys. Rev.* **181**, 1297 (1969).

[158] N. Schwentner, E.-E. Koch, and J. Jortner, *Excitonic excitations in condensed rare gases*, Springer-Verlag, Berlin (1985).

[159] R. Opila and R. Gomer, *Surface Science* **112**, 1 (1981).

[160] T.-C. Chiang, G. Kaindl, and T. Mandel, *Phys. Rev. B* **33**, 695 (1986).

[161] T. Mandel, M. Domke, and G. Kaindl, *Surface Science* **197**, 81 (1988).

[162] G. Kaindl, T. C. Chiang, D. E. Eastman, and F. J. Himpsel, *Phys. Rev. Lett.* **45**, 1808 (1980).

[163] G. Kaindl, T. C. Chiang, and T. Mandel, *Phys. Rev. B* **28**, 3612 (1983).

[164] B. Kessler, A. Eyers, K. Horn, N. Müller, B. Schmiedeskamp, G. Schönhense, and U. Heinzmann, *Phys. Rev. Lett.* **59**, 331 (1987).

[165] T. Engel, P. Bornemann, and E. Bauer, *Surface Science* **81**, 252 (1979).

[166] A. Ausmees, S. J. Osborne, R. Moberg, S. Svensson, S. Aksela, O.-P. Sairanen, A. Kivimäki, A. Naves de Brito, E. Nõmmiste, J. Jauhiainen, and H. Aksela, *Phys. Rev. A* **51**, 855 (1995).

[167] L. Verlet, *Phys. Rev.* **159**, 98 (1967).

[168] R. Opila and R. Gomer, *Surface Science* **127**, 569 (1983).

[169] A. S. Kheifets and I. A. Ivanov, *Phys. Rev. Lett.* **105**, 233002 (2010).

[170] S. Nagele, R. Pazourek, J. Feist, K. Doblhoff-Dier, C. Lemell, K. Tőkési, and J. Burgdörfer, *J. Phys. B* **44**, 081001 (2011).

[171] M. Ivanov and O. Smirnova, *Phys. Rev. Lett.* **107**, 213605 (2011).

[172] C.-H. Zhang and U. Thumm, *Phys. Rev. A* **82**, 043405 (2010).

[173] M. Tchaplyguine, R. R. Marinho, M. Gisselbrecht, J. Schulz, N. Mårtensson, S. L. Sorensen, A. Naves de Brito, R. Feifel, G. Öhrwall, M. Lundwall, S. Svensson, and O. Björneholm, *J. Chem. Phys.* **120**, 345 (2004).

Bibliography

i want morebooks!

Buy your books fast and straightforward online - at one of world's fastest growing online book stores! Environmentally sound due to Print-on-Demand technologies.

Buy your books online at
www.get-morebooks.com

Kaufen Sie Ihre Bücher schnell und unkompliziert online – auf einer der am schnellsten wachsenden Buchhandelsplattformen weltweit! Dank Print-On-Demand umwelt- und ressourcenschonend produziert.

Bücher schneller online kaufen
www.morebooks.de

 VDM Verlagsservicegesellschaft mbH
Heinrich-Böcking-Str. 6-8   Telefon: +49 681 3720 174   info@vdm-vsg.de
D - 66121 Saarbrücken   Telefax: +49 681 3720 1749 www.vdm-vsg.de

Printed by Books on Demand GmbH, Norderstedt / Germany